Milestones in Graph Theory

A Century of Progress

AMS/MAA | SPECTRUM

VOL **108**

Milestones in Graph Theory

A Century of Progress

Lowell W. Beineke

Bjarne Toft

Robin J. Wilson

2020 *Mathematics Subject Classification.* Primary 01A60, 05C10, 05C15, 05C20, 05C35, 05C45, 05C50, 05C55, 05C70, 05C85.

Library of Congress Cataloging-in-Publication Data

Cataloging-in-Publication Data has been applied for by the AMS.
See http://www.loc.gov/publish/cip/.

Contents

Preface

Unlike the development of many areas of mathematics, that of graph theory was largely a 20th-century phenomenon. Although the subject had its origins in the 18th and 19th centuries, it did not come to fruition until the following century. From the Königsberg bridges problem onwards, many of the earlier explorations into (and utilizations of) graphs and networks focused on applications, either to puzzles and games, or to scientific questions and practical problems.

That early portion of the subject's history has been fully documented in N. L. Biggs, E. K. Lloyd, and R. J. Wilson's *Graph Theory 1736–1936*, which concludes with the publication of the first textbook on graph theory (D. König's *Theorie der endlichen und unendlichen Graphen*) and whose story "from Königsberg to König's book" is told through a narrative on the topics and the authors, interspersed with excerpts from their papers. Another historical book, by Robin Wilson, John J. Watkins, and David J. Parks, is *Graph Theory in America*, which chronicles the American story from 1876 (when J. J. Sylvester became the first mathematics professor at Johns Hopkins University) to 1976 (when Kenneth Appel and Wolfgang Haken proved the four-color theorem). The aim of the present book is to add to the historical journey by detailing some of the routes that were traveled, focusing on some of the "milestones" that were encountered in various countries along the way. Several of these milestones are now regarded as classics.

By the middle of the 20th century, significant advances in graph theory had been made throughout Europe and the United States. Many mathematicians who contributed to the story are remembered only for their work in graph theory, while others, such as George Birkhoff, Kazimierz Kuratowski, and Hassler Whitney, are recognized for their work in other fields and it was relatively recently that their substantial contributions to graph theory also came to be appreciated.

An early criticism of the subject was that it lacked theory, being primarily a collection of elementary results, albeit useful, interesting, and often entertaining. But while an important feature of graph theory has continued to be a strong problem-solving thread, deep theoretical results began to be developed alongside this, with notable examples coming from the series of pioneering papers by Whitney in the early 1930s, the work of Bill Tutte in the 1940s and 1950s, the theoretical advances by Hungarian mathematicians, among them Paul Turán, Alfréd Renyi, Tibor Gallai, Paul Erdős, Béla Bollobás, Endre Szemerédi and László Lovász, and the graph minors project of Neil Robertson and Paul Seymour and their colleagues around the end of the 20th century.

Furthermore, since the 1950s the theory has developed significantly through the influence of computer science and algorithmic mathematics—for example, through the classic question "Does $P = NP$?" and its connections with such problems as deciding whether a given graph is 3-colorable or Hamiltonian. Algorithmic graph theory and optimization were developed by Jack Edmonds, Richard Karp, Stephen Cook, Robert Tarjan, László Lovász, Alexander Schrijver, András Frank, Éva Tardos, and others.

While writing this book, we examined hundreds of original papers in graph theory and related topics, as we identified the various directions in which the subject developed and noted milestones along the way. Although most of our discussions focus on the 20th century, we have felt free to expand the timescale slightly to what historians sometimes call a "long century", allowing also the pioneering contributions of Arthur Cayley, Percy Heawood, and Julius Petersen in the 1880s and 1890s and some recent work from the present century.

After an introductory chapter that sets the scene with the legacy from the 18th and 19th centuries, this book is organized thematically in seven further chapters, each surveying the developments in a specified area. We hope that you will thereby be enabled to appreciate the struggles of the early pioneers, experience the rapid expansion of the subject through the 1960s and 1970s, and see how it increasingly became part of mainstream mathematics in the 1980s and 1990s, continuing to this day and into the future. A timeline provides a guide to the development of the subject.

We are deeply grateful for the inclusive and open atmosphere that we have always enjoyed in the graph theory community. The editors of the *Spectrum* series and the AMS editorial and production staff have provided invaluable assistance—in particular we thank Stephen Kennedy, Ina Mette, John Lorch, Thomas Sharland, Abigail Lawson, and the anonymous referees.

This book is not written only for graph theorists, but should also be accessible to anyone with a sufficient level of mathematical experience and sophistication, including students, teachers, and others interested in mathematics and its history and development. Supplementary resources are listed at the end of each chapter for those wishing to explore the topics in greater depth. We hope that it will provide many hours of profitable study and enjoyment.

Lowell Beineke, Bjarne Toft, and Robin Wilson

Graph Theory Timeline

In this timeline we list some of the important publications and events in the history of graph theory, from the 18th century to the present day.

1735   Euler solves the Königsberg bridges problem
1750   Euler states his "polyhedron formula" $F + V = E + 2$
1759   Euler discusses the knight's tour on a chessboard problem
1794   Legendre proves Euler's polyhedron formula
1812   Lhuilier proves and extends Euler's formula
1813   Cauchy reformulates and proves Euler's formula for planar graphs
1847   Kirchhoff writes on electrical networks and introduces spanning trees
1849   Listing writes on path-tracing problems having introduced the word "topology"
1852   Guthrie proposes, and De Morgan propagates, the four-color problem for maps
1856   Kirkman and Hamilton trace "Hamiltonian cycles" on polyhedra
1857   Cayley writes his first paper on trees
1861   Listing continues investigations into topology by discussing "spatial complexes"
1875   Cayley enumerates certain chemical molecules
1878   Cayley revives the four-color problem at a meeting of the London Mathematical Society
Cayley introduces "Cayley graphs"
Sylvester writes on chemical molecules and introduces the word "graph"
1879   Cayley proves that the four-color problem can be reduced to cubic maps
Kempe proposes a proof of the four-color theorem
1880   Tait reformulates the four-color problem in terms of coloring the edges of cubic graphs
1881   Cayley summarizes his results on the enumeration of trees
1889   Cayley presents his n^{n-2} result on the enumeration of labeled trees

1890   Heawood points out the error in Kempe's proof of the four-color theorem, proves the five-color theorem, and discusses the coloring of maps on surfaces

1891   Heffter explores the coloring of maps on orientable surfaces

Petersen discusses the factorization of regular graphs

1892   Rouse Ball presents the graph associated with the Königsberg bridges problem

1895   Tarry presents a method for tracing a maze

1898   Heawood explores the congruences arising from coloring cubic maps

Petersen introduces the "Petersen graph"

1904   Wernicke introduces a new unavoidable set for the coloring of maps

1907   Dehn and Heegaard write on "Analysis situs" for the German Encyclopedia of Mathematics

Wythoff solves an extremal problem of Mantel on graphs without triangles

1910   Tietze explores the coloring of maps on nonorientable surfaces

1912   Birkhoff introduces chromatic polynomials

Veblen writes on the algebra of graphs and poses the four-color problem in these terms

1913   Birkhoff discusses reducible configurations in maps and introduces the "Birkhoff diamond"

1916   König discusses bipartite graphs and the coloring of their edges

1918   Prüfer proves Cayley's n^{n-2} result for labeled trees

1920s   Errera writes several papers on map coloring

1922   Franklin writes on map coloring and proves that all maps with up to 25 countries can be 4-colored

1926   Borůvka solves the minimum connector problem on minimum-length spanning trees

Reynolds proves that all maps with up to 27 countries can be 4-colored

Sainte-Lagüe writes an introductory book *Les réseaux (ou graphes)* (English version, 2020)

1927   Menger investigates the connectivity of graphs

Redfield discusses the enumeration of graphs with symmetry

1930   Jarník solves the minimum connector problem

Kuratowski obtains a forbidden-subgraph characterization of planar graphs

Ramsey writes on a problem in set theory, laying the groundwork for "Ramsey graph theory"

1931   König writes on graphs and matrices

Whitney writes on the duality of planar graphs

Whitney writes on the coloring of graphs and chromatic polynomials

1932   Whitney writes on the connectivity of graphs
1934   Franklin proves that all maps on a Klein bottle can be colored
       with six colors
       Rédei investigates Hamiltonian paths in tournaments
1935   Hall proves "Hall's theorem" on representatives of subsets
       Klein (Esther) poses an extremal problem in geometry
       Whitney writes a pioneering paper on matroids
1936   König publishes *Theorie der endlichen und unendlichen
       Graphen* (English version, 1990)
1937   Wagner writes on a property of plane complexes
       Mac Lane obtains a new condition for a graph to be planar
       Pólya writes a groundbreaking paper on graph enumeration
1938   Frucht proves that every finite group is the automorphism
       group of some graph
1940   Brooks, Smith, Stone, and Tutte solve the problem of "squaring
       the square"
       Lebesgue finds new unavoidable sets for map coloring
       Turán obtains a bound on the crossing number of complete
       bipartite graphs
       Winn proves that all maps with up to 35 countries can be 4-
       colored
1941   Brooks proves an important result on the vertex-coloring of
       graphs
       Turán solves the central problem in extremal graph theory
       Hitchcock formulates the transportation problem
1943   Hadwiger poses his conjecture on the contraction of graphs
1946   Birkhoff and Lewis publish a major paper on chromatic poly-
       nomials
       Tutte presents a counter-example to Tate's conjecture on
       Hamiltonian cycles in cubic polyhedra
1947   Erdős invents the probabilistic method and obtains a bound
       for a problem in extremal graph theory
       Tutte presents a condition for a graph to have a 1-factor
       Tutte introduces the "Tutte polynomial" of a graph
1948   Fáry proves that every planar graph can be drawn in a plane
       with straight line segments (proved earlier by Wagner, but un-
       noticed)
       Otter investigates the enumeration of trees
1949   Shannon publishes an upper bound for edge-colorings of
       multigraphs
1950   Belck publishes a condition for a graph to have an r-factor and
       gives the first purely graph-theoretic proof of Tutte's 1-factor
       theorem
1952   Dirac presents a condition for a graph to be Hamiltonian and
       obtains results on critical graphs for coloring
       Ringel proves the Heawood conjecture for coloring maps on
       nonorientable surfaces
       Tutte proves a general factor theorem

1967   Edmonds conjectures that there is no polynomial algorithm for solving the traveling salesman problem, and thus that $P \neq NP$
Ore publishes *The Four-Color Problem*
Rosa presents a systematic study of graph labelings
First Chapel Hill (North Carolina, USA) conference on combinatorial mathematics

1968   Beineke presents a forbidden subgraphs characterization of line graphs
First Kalamazoo (Michigan) quadrennial conference on graph theory
Moon publishes *Topics on Tournaments*
Ringel and Youngs solve the Heawood conjecture for coloring maps on orientable surfaces

1969   Harary publishes *Graph Theory*
Heesch writes *Untersuchungen zum Vierfarbenproblem*
First British Combinatorial Conference, in Oxford

1970   Berge publishes *Graphes et hypergraphes* (English version, 1973)
First Southeastern Conference on Combinatorics, Graph Theory, and Computing, in Baton Rouge (Louisiana, USA)

1971   Biggs and Smith write on trivalent graphs
Cook writes a classic paper on complexity, introducing NP-completeness
Discrete Mathematics is launched

1972   Behzad and Chartrand write *Introduction to the Theory of Graphs*
Lovász characterizes perfect graphs and proves the weak perfect graph conjecture
Lovász proves a result on reconstructing graphs
Wilson (Robin) writes *Introduction to Graph Theory*

1973   Harary and Palmer write *Graphical Enumeration*
Pinsker introduces expander graphs and concentrators

1974   Hopcroft and Tarjan discuss the complexity of planarity testing
Ringel publishes *Map Color Theorem*

1975   Fulkerson edits *Studies in Graph Theory, I and II*
Karp surveys the complexity of graph problems
Wilson (Richard) investigates the decomposition of complete graphs into isomorphic subgraphs

1976   Appel and Haken prove the four-color theorem
Biggs, Lloyd, and Wilson publish *Graph Theory 1736–1936* on the history of graph theory
Bondy and Chvátal investigate Hamiltonian graphs
Bondy and Murty publish *Graph Theory with Applications*

1977	Appel, Haken, and Koch publish their proof of the four-color theorem Cockayne and Hedetniemi write on domination in graphs *Journal of Graph Theory* is launched
1978	Beineke and Wilson publish the first volume of *Selected Topics in Graph Theory* Bollobás publishes *Extremal Graph Theory* Glover and Huneke present the forbidden subgraphs and minors for graphs drawn on a projective plane Szemerédi obtains the regularity lemma and investigates random graphs, in work that contributes to his 2012 Abel Prize
1979	Garey and Johnson publish *Computers and Intractability: A Guide to Theory of NP-completeness* Lovász writes *Combinatorial Problems and Exercises*
1980	Cameron writes on graphs with a given automorphism group
1981	*Combinatorica* is launched
1982	Luks investigates the graph isomorphism problem
1983	Robertson and Seymour begin their extended graph minors project, with twenty papers, from tree-width (1983) to a proof of a conjecture of Wagner (2004) Shearer obtains an upper bound for the Ramsey numbers $r(3, l)$
1984	Alon and Milman write on eigenvalues and expanders Robertson and Seymour prove a Kuratowski-type theorem for general surfaces
1985	Bollobás publishes *Random Graphs* For the min cost flow problem, Éva Tardos presents a polynomial algorithm with running time independent of costs and capacities
1990s	Several graph databases are developed to store and retrieve information (graph analytics), among them social graphs, interest graphs, and the web graph (Facebook, Twitter, Amazon, Google)
1990	The Institute of Combinatorics and its Applications is founded Bollobás proves that almost all graphs are reconstructible from only three vertex-deleted subgraphs Robertson and Seymour prove a Kuratowski-type theorem for general surfaces
1990–91	Fleischner publishes two volumes on *Eulerian Graphs and Related Topics*
1994	Thomassen writes on list-colorings, proving that every planar graph is 5-choosable
1995	Kim determines the asymptotic behavior of the Ramsey numbers $r(3, l)$ The two-volume *Handbook of Combinatorics* is published with editors Graham, Grötschel, and Lovász

1996   Diestel publishes *Graphentheorie* (English version, 1997)
        Robertson, Sanders, Seymour, and Thomas present a more
        systematic proof of the four-color theorem
        West publishes *Introduction to Graph Theory*
1998   Haynes, Hedeniemi, and Slater publish two books on domi-
        nation in graphs
        Page and Brin introduce *PageRank*, the graph search tool of
        Google
2001   Mohar and Thomassen publish *Graphs on Surfaces*
2002   Bang-Jensen and Gutin publish *Digraphs: Theory, Algorithms
        and Applications*
        Molloy and Reed publish *Graph Colouring and the Probabilis-
        tic Method*
2003   Schrijver publishes the three-volume *Combinatorial Opti-
        mization, Polyhedra and Efficiency*
2004   Gonthier provides a formal computer check on the proof of
        the four-color theorem
        Handbook of Graph Theory appears with editors Gross and
        Yellen
2006   Chudnovsky, Robertson, Seymour, and Thomas publish a
        proof of the strong perfect graph conjecture
2008   Steinberger uses only D-reducible configurations to prove the
        four-color theorem
2012   Lovász writes *Large Networks and Graph Limits*
        Szemerédi is awarded the Abel Prize
2021   Lovász and Wigderson are awarded the Abel Prize
2024   Mattheus and Verstraete determine the asymptotic behavior
        of the Ramsey numbers $r(4, l)$

1

Setting the Scene

This book focuses mainly on the development of graph theory in the 20th century. In order to set the scene, we briefly describe here some major achievements from earlier times. Most of these topics are developed in later chapters. We begin with a few basic definitions.

1.1 Introduction

A *graph G* consists of a finite set $V(G)$ of *vertices* and a finite set $E(G)$ of *edges*, where each edge joins two vertices (see Figure 1.1). Two or more edges joining the same pair of vertices are called *multiple edges* and the graph is then sometimes called a *multigraph*. An edge joining a vertex to itself is a *loop*. A graph with no loops or multiple edges is a *simple graph*, and most of the graphs in this book are simple unless the context implies otherwise.

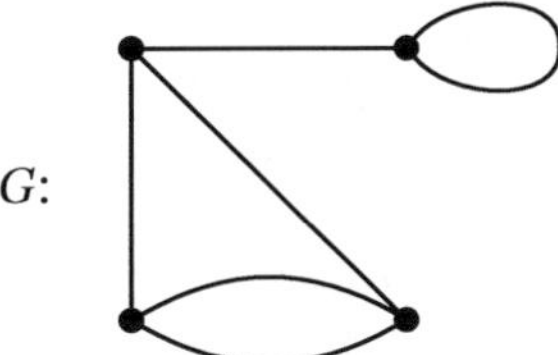

Figure 1.1. A graph G with four vertices and six edges

A *digraph D*, sometimes called a *directed graph*, is the result of giving each edge of a graph a direction, then calling it an *arc*. Figure 1.2 is an illustration of a digraph.

A *subgraph* of a graph G is a graph H with $V(H) \subseteq V(G)$ and $E(H) \subseteq E(G)$. If an edge e joins vertices v and w, then we often write $e = vw$. We also say that v and w are *adjacent* and that v and w are *incident* with e. The *degree* deg v of a vertex v is the number of edges incident with v (with loops counted twice); thus, in Figure 1.1, each vertex has degree 3. If all vertices of a graph have the same degree, the graph is called

1

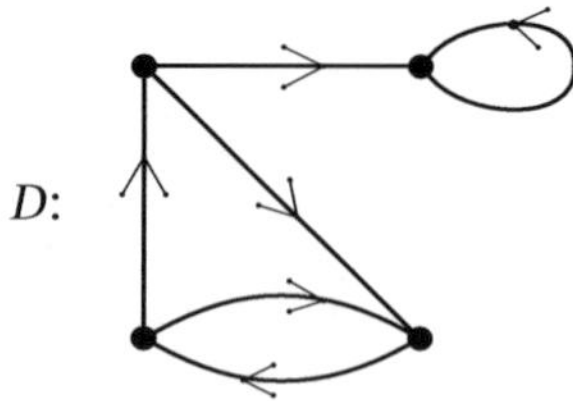

Figure 1.2. A digraph D with four vertices and six arcs

regular and is *k-regular* if that degree is k. A 3-regular graph is sometimes called *cubic*, and an example is the so-called *Petersen graph* (see Figure 1.3).

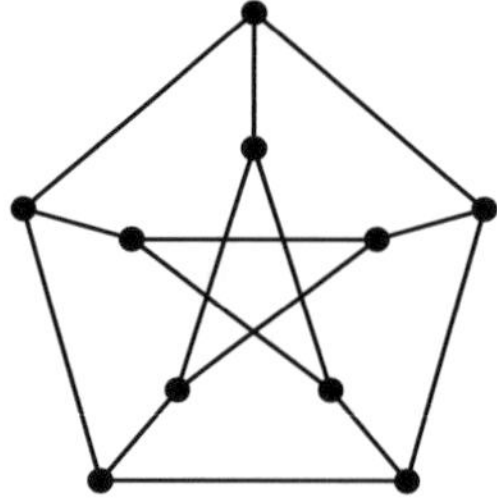

Figure 1.3. The Petersen graph

A *walk of length k* in a graph is a sequence $v_0, e_1, v_1, e_2, \ldots, e_k, v_k$, of vertices and edges beginning and ending at vertices and with each edge e_i joining vertices v_{i-1} and v_i. It is called a *closed walk* if $v_0 = v_k$. A *path* is a walk in which no vertex is repeated, a *cycle* is a closed walk with no edge repeated and no vertex repeated except the first and last. A graph is *connected* if there is a path connecting each pair of vertices, and is *disconnected* otherwise.

A graph with n vertices with each pair joined by exactly one edge is called a *complete graph* and is denoted K_n. The *cycle graph* C_n is the graph consisting of a single cycle with n vertices and n edges. A graph whose vertices are split into two disjoint sets with no edges joining two vertices in the same set is called *bipartite*. The two sets are called *partite sets,* and if the numbers of vertices in them are r and s, the graph is an $r \times s$ *bipartite graph*. If each pair of vertices from different sets is joined by an edge, this graph is the *complete bipartite graph* $K_{r,s}$. Figure 1.4 shows the graphs K_5 and $K_{3,3}$; we shall meet them again in Chapter 3.

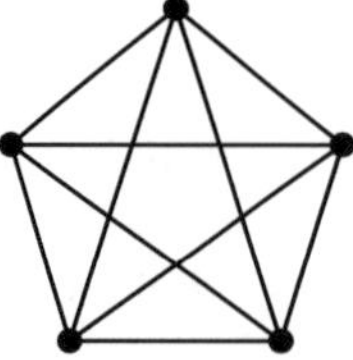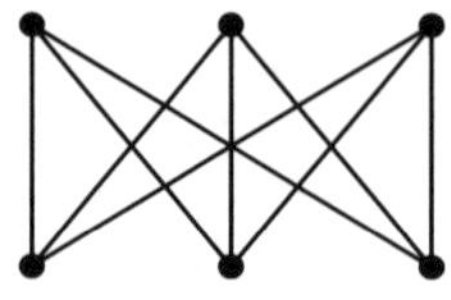

Figure 1.4. The graphs K_5 and $K_{3,3}$

1.2　Euler and the Königsberg bridges

In 1735, the Swiss mathematician Leonhard Euler presented to his colleagues in St Petersburg his "solution of a problem relating to the geometry of position".[1] The medieval city of Königsberg in Eastern Prussia consisted of four regions linked by seven bridges (see Figure 1.5), and the problem was to find a route that crosses each of the seven bridges exactly once before returning to the starting point.

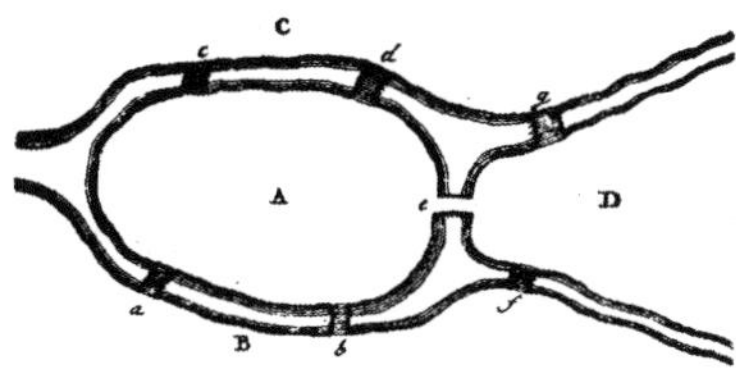

Figure 1.5.　The seven bridges of Königsberg

Euler extended the problem to general arrangements of regions and bridges, and showed that such a route can exist only when the number of bridges out of each region is even; the Königsberg problem has no solution because there are five bridges from region A and three bridges from regions B, C, and D.

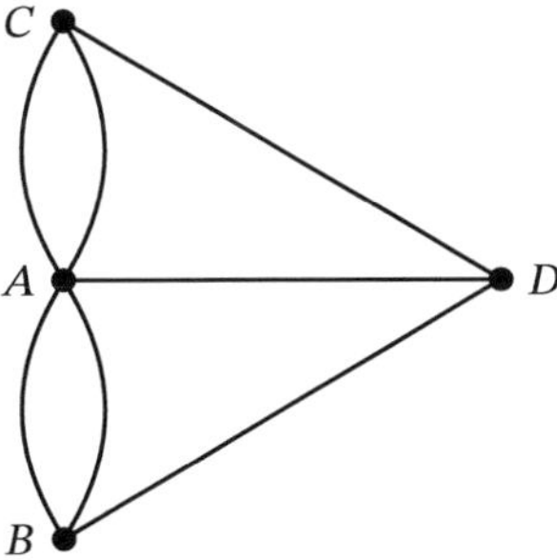

Figure 1.6.　The Königsberg graph

It has often been asserted that Euler solved the problem by drawing the multigraph in Figure 1.6, with four vertices (the regions) and seven edges (the bridges), but this is not the case. The first appearance of this graph seems to have been in 1892 when W. W. Rouse Ball, in his popular book *Mathematical Recreations and Problems*,[2] recast the Königsberg problem as that of deciding whether this graph can be drawn in a single stroke, repeating no edge and returning to the starting vertex. Graphs where this is possible are now called *Eulerian graphs* and can be characterized as follows:

*A graph is Eulerian if and only if it is connected
and the degree of each vertex is even.*

For the Königsberg graph, the degrees of the vertices are odd, and so the graph is not Eulerian.

1.3 Hamiltonian cycles

Another type of graph problem involves finding a cycle that passes exactly once through every vertex of a graph before returning to the start. Such a cycle is called a *Hamiltonian cycle* and the graph is a *Hamiltonian graph*, named after the Irishman William Rowan Hamilton.

In 1856, Hamiltonian showed that the graph of the dodecahedron has such cycles—an example appears in Figure 1.7, where the vertices are labeled by the twenty consonants and the cycle visits them in alphabetical order.[3] Hamilton subsequently turned the idea into a puzzle, the *Icosian game*, where the vertices represent twenty cities from around the world (Brussels, Canton,..., Zanzibar). Marketed as *A Voyage around the World*, the aim was to visit every city just once, according to certain specified instructions.

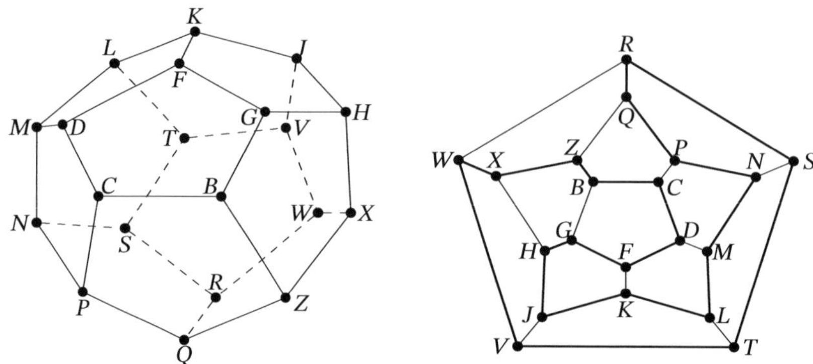

Figure 1.7. A 3-dimensional drawing of a dodeca-hedron and a Hamiltonian cycle in its graph

Hamilton was not the first to study cycles on polyhedra; similar investigations had already been made by the English clergyman Thomas P. Kirkman, who considered various examples and who really deserves the credit for studying them.[4] Unlike the Eulerian problem, no conditions that are both necessary and sufficient are known for the existence of a Hamiltonian cycle in a general graph. We revisit Hamiltonian graphs in Chapter 5.

A much earlier cycle problem is the *knight's tour problem*:

> *Can a knight visit all of the 64 squares of a chessboard just once*
> *before returning to its starting point?*

(Here, a knight moves two squares in one direction and one square in a perpendicular direction.) We see the connection with Hamiltonian graphs by regarding the squares as vertices of a graph, with a pair of vertices joined by an edge whenever the corresponding squares are linked by a single knight's move. A solution by Euler appears in Figure 1.8.

1.4 Euler's polyhedron formula

The Greeks were familiar with the five regular solids (the tetrahedron, cube, octahedron, dodecahedron, and icosahedron) and other polyhedra, but there is no evidence

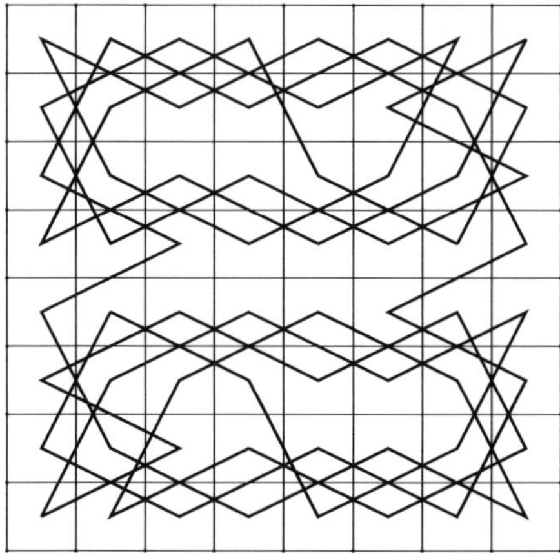

Figure 1.8. Euler's knight's tour on a chessboard

that they knew the simple result that relates the numbers of vertices, edges, and faces:

$$\text{(number of faces)} + \text{(number of vertices)} = \text{(number of edges)} + 2.$$

For example, a dodecahedron has 12 faces, 20 vertices, and 30 edges, and $12 + 20 = 30 + 2$.

While investigating polyhedra, Euler introduced the concept of an "edge", and in 1750 he stated this result in a letter to Christian Goldbach. He also verified the formula for several polyhedra, but his attempts at a proof were deficient. The first valid demonstration was given by A.-M. Legendre in 1794; unlike Euler's attempts, it used metrical ideas.[5]

A graph is *planar* if it can be drawn in the plane without edges crossing; for example, the graph in Figure 1.9 is planar. Such a drawing divides the plane into regions called *faces*, and Euler's formula still holds, provided that we include the "infinite" (unbounded) face. In 1813, by projecting polyhedra onto a plane, the French mathematician Augustin-Louis Cauchy obtained a nonmetrical proof of Euler's formula for graphs drawn in the plane.[6]

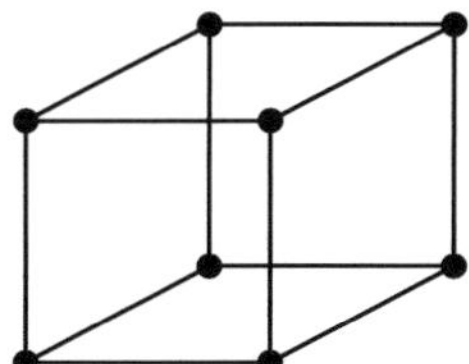
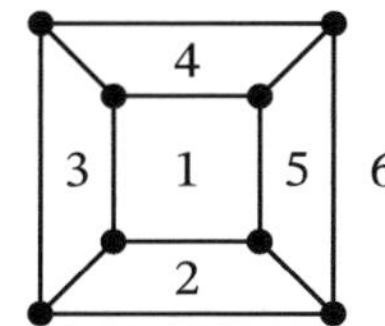

Figure 1.9. A planar graph and a plane drawing with six faces

Around the same time, Simon-Antoine-Jean Lhuilier (or l'Huilier) discovered three types of polyhedra for which Euler's formula fails.[7] These included polyhedra with a "tunnel" through them, and for such polyhedra he obtained the formula

$$\text{(number of vertices)} - \text{(number of edges)} + \text{(number of faces)} = 0.$$

He then extended his discussion to prove that if a polyhedron has g tunnels, then

$$\text{(number of vertices)} - \text{(number of edges)} + \text{(number of faces)} = 2 - 2g.$$

This number g depends only on the surface, and not on the polyhedron drawn on it, as we explain in Chapter 3.

Lhuilier's results were the starting point for an extensive investigation by Johann Benedict Listing, titled *Der Census räumliche Complexe* (The Census of Spatial Complexes).[8] These "complexes" are built from simpler pieces, and Listing investigated how their topological properties affect the above extension of Euler's formula. His ideas were later developed by the French mathematician Henri Poincaré in his papers that laid the foundations of algebraic topology. Like Listing, Poincaré developed a method for constructing complexes from basic "cells", such as 0-cells (vertices) and 1-cells (edges). To fit these cells together, he adapted a technique of Gustav Robert Kirchhoff from the theory of electrical networks, replacing sets of linear equations by matrices.[9] These matrices could then be studied from an algebraic point of view, as we illustrate in Chapter 4.

1.5 Trees

We are all familiar with the idea of a family tree. Mathematically, a *tree* is a connected graph with no cycles, and Figure 1.10 illustrates all the trees with six vertices and five edges. In general, every tree with n vertices has exactly $n - 1$ edges.

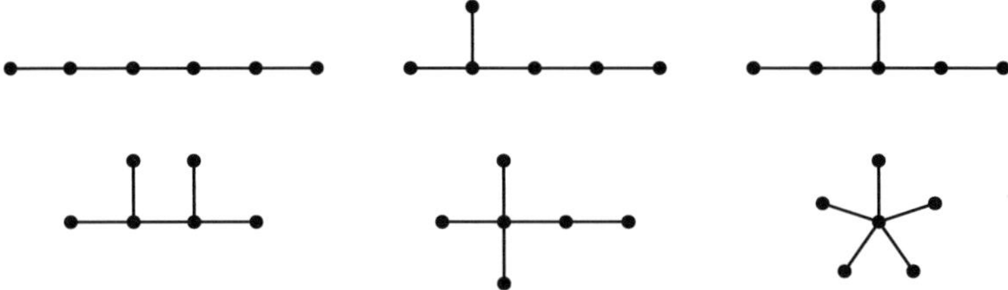

Figure 1.10. The six trees with six vertices

The concept of a tree arose implicitly in 1847 in the above-mentioned work of Kirchhoff, who used them while calculating currents in an electrical network. Ten years later, inspired by a problem on the differential calculus that had been raised by his friend James Joseph Sylvester, the Cambridge mathematician Arthur Cayley became interested in tree-counting problems.[10] He also became involved with the related problem of enumerating certain chemical molecules with a tree structure, such as the paraffins (alkanes) with chemical formula C_nH_{2n+2}—examples are regular butane (or n-butane) and 2-methyl propane, both of which have formula C_4H_{10} (see Figure 1.11). Such problems can be reduced to the counting of trees in which every vertex has degree 4 (for carbon) or degree 1 (for hydrogen).

Sylvester was also interested in chemical molecules and in their diagrammatic representation, and in 1878 he introduced the word *graph* for the "chemico-graphical images" just illustrated.[11] He also believed there to be a link between them and the diagrams that arose in the algebraic theory of invariants (see Figure 1.12), and wrote a lengthy paper on the subject in the *American Journal of Mathematics* which he had just helped to found.[12]

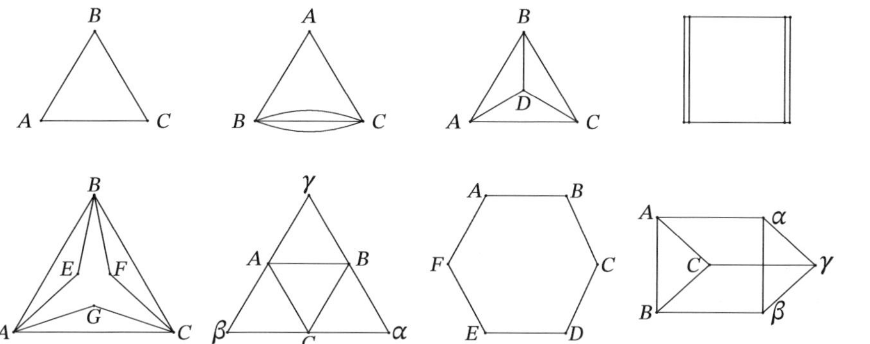

Figure 1.11. The molecules with chemical formula C_4H_{10}

Figure 1.12. Some of Sylvester's chemical graphs

1.6 The four-color problem

The earliest known reference to the "four-color problem" on the coloring of maps was in a letter sent to Hamilton in 1852 by Augustus De Morgan in London.[13] In his letter, De Morgan described how one of his students had asked whether every map can be colored with only four colors so that neighboring countries are colored differently, as in Figure 1.13; the student, Frederick Guthrie, later claimed that it was actually his brother Francis who first raised the problem while coloring a map of England.[14] De Morgan observed that four colors are needed whenever the map includes four mutually neighboring countries, but that four colors may be required even when this is not the case.

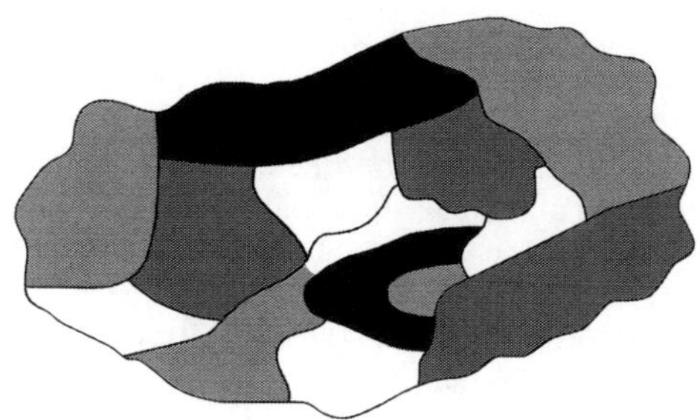

Figure 1.13. A map that requires four colors

It was not until after De Morgan's death in 1871 that progress was made in proving that four colors suffice for all maps drawn on a plane. In 1878, at a meeting of the London Mathematical Society, Cayley enquired whether the problem had been solved, and soon afterwards proved that we can assume the map to be *cubic*, with exactly three countries meeting at each point of intersection. He also showed that it makes no difference whether or not the external (unbounded) region is colored.

In 1879, Alfred B. Kempe, a London barrister who had studied with Cayley at Cambridge and had attended the London meeting, presented an argument that all maps can be colored with four colors; it was subsequently published in the *American Journal of Mathematics*.[15] Eleven years later, Percy Heawood of Durham pointed out a major error in Kempe's "proof" but managed to salvage enough from Kempe's argument to prove that every map can be colored with five colors—itself a remarkable result.[16] Although Kempe's argument contained a fatal flaw, it included important ideas that were to feature in later attempts on the problem, as we shall see in Chapter 2.

Further Reading

An extended version of this chapter appears as Robin Wilson's chapter on "Early graph theory" in

> *Combinatorics: Ancient & Modern* (ed. Robin J. Wilson and John J. Watkins), Oxford (2013); paperback edn. (2015).

The history of graph theory is comprehensively covered in

> N. L. Biggs, E. K. Lloyd, and R. J. Wilson, *Graph Theory 1736–1936*, Oxford (1976); paperback edn. (1996).

> Robin Wilson, John J. Watkins, and David J. Parks, *Graph Theory in America*, Princeton Univ. Press (2023).

Introductory books on graph theory include

> Daniel A. Marcus, *Graph Theory: A Problem Oriented Approach*, MAA (2008).

> Oystein Ore, *Graphs and Their Uses* (updated edn. by R. Wilson), MAA (1990).

> Robin J. Wilson, *Introduction to Graph Theory* (5th edn.), Prentice-Hall (2010).

Notes and References

[1] L. Euler, Solutio problematis ad geometriam situs pertinentis, *Comm. Acad. Sci. Imp. Petropol.* 8 (1736), 128–140 = *Opera Omnia* (1) 7, 1–10; although dated 1736, it was not published until 1741. In this paper Euler presented this problem as an example of "the geometry of position" in response to a desire by Gottfried Leibniz for a type of geometry that involved no metrical ideas, such as length and angle. Euler also wrote several letters on the Königsberg bridges problem; for details, see H. Sachs, M. Stiebitz, and R. J. Wilson, An historical note: Euler's Königsberg letters, *J. Graph Theory* 12 (1988), 133–139; Brian Hopkins and Robin Wilson, The truth about Königsberg, *College Math. J.* 35 (2004), 198–207.

[2] W. W. Rouse Ball, *Mathematical Recreations and Problems of Past and Present Times*, Macmillan (1892). In later editions, the title was changed to *Mathematical Recreations and Essays*.

[3] W. R. Hamilton, Memorandum respecting a new system of roots of unity, *Phil. Mag.* (4) 12 (1856), 446 = *Math. Papers*, Vol. 3, 610. In this note, Hamilton showed how a particular algebraic

system of complex roots of 1 could be interpreted in terms of cycles of faces of an icosahedron, or equivalently of vertices of a dodecahedron.

[4]T. P. Kirkman, On the representation of polyedra, *Phil. Trans. Roy. Soc. London* 146 (1856), 413–418. Unlike Hamilton, Kirkman investigated cycles on several polyhedra and gave examples of when such cycles do or do not exist.

[5]Euler's letter to Goldbach is in his *Opera Omnia* (4), I, letter 863, and his attempted proof appears in his article, Elementa doctrinae solidorum, *Novi Comm. Acad. Sci. Imp. Petropol.* 4 (1752–53), 109–140 = *Opera Omnia* (1), Vol. 26, 72–93. Legendre's metrical proof appears in his *Eléments de géométrie*, Firmin Didot, Paris (1794). The polyhedron formula has often been credited to René Descartes who did not have the concept of an "edge" and never made the deduction.

[6]A.-L. Cauchy, Recherches sur les polyèdres—premier mémoire, *J. École Polytech.* 9 (Cah. 16) (1813), 68–86 = *Oeuvres* (2), Vol. 1, 7–25. This was an early work of Cauchy, who later became known mainly for his major contributions to real and complex analysis.

[7]Lhuilier, Démonstration immédiate d'un théorème fondamental d'Euler sur les polyhèdres, et exceptions dont ce théorème est susceptible, *Mém. Acad. Imp. Sci. St. Pétersb.* 4 (1811), 271–301. In the same paper, Lhuilier explained why there are only five regular polyhedra.

[8]J. B. Listing, Der Census räumliche Complexe oder Veralgemainerung des Euler'schen Satzes von den Polyëdern, *Abh. K. Ges. Wiss. Göttingen Math. Cl.* 10 (1861–62), 97–182. This paper can be thought of as one of the seminal papers in *topology*, a word that he had coined in a letter in 1838. It first appeared in print in 1847, in his book *Vorstudien zur Topologie* (Introductory Studies in Topology).

[9]G. R. Kirchhoff, Über die Auflösung der Gleichungen, auf welche man bei der Untersuchung der linearen Vertheilung galvernischer Ströme geführt wird, *Ann. Phys. Chem.* 72 (1847), 497–508 = *Gesammelte Abhandlungen*, 22–33.

[10]A. Cayley, On the theory of the analytical forms called trees, *Phil. Mag.* (4) 13 (1857), 172–176 = *Math. Papers*, Vol. 3, 242–246. Cayley followed this by a second paper in 1859, and by papers on the enumeration of chemical molecules in the 1870s.

[11]J. J. Sylvester, Chemistry and algebra, *Nature* 17 (1877–78) = *Math. Papers*, Vol. 3, 103–104.

[12]J. J. Sylvester, On an application of the new atomic theory to the graphical representation of the invariants and covariants of binary quantics—with three appendices, *Amer. J. Math.* 1 (1878), 64–125 = *Math. Papers*, Vol. 3, 148–206. Here, in a remarkable paragraph, Sylvester described how he found graphs to be a useful tool for demonstrating links between algebra and chemistry: "Casting about, as I lay awake one night, to discover some means of conveying an intelligible conception of objects of modern algebra to a mixed society, mainly composed of physicists, chemists and biologists, interspersed only with a few mathematicians, to which I stood engaged to give some account of my recent researches ... I was agreeably surprised to find, of a sudden, distinctly pictured on my mental retina, a chemico-graphical image serving to embody and illustrate the relations ..."

[13]For more information, see Robin Wilson, *Four Colors Suffice* (revised color edn.), Princeton Univ. Press (2014). This book was first published (in black and white) by Allan Lane in 2002 to celebrate the 150th anniversary of Francis Guthrie's posing of the four-color problem.

[14]Frederick Guthrie, Note on the colouring of maps, *Proc. Roy. Soc. Edinburgh* 10 (1880), 727–728.

[15]A. B. Kempe, On the geographical problem of the four colours, *Amer. J. Math.* 2 (1879), 193–200. This paper contained Kempe's incorrect proof of the four-color theorem (see Chapter 2).

[16]P. J. Heawood, Map-colour theorem, *Quart. J. Pure Appl. Math.* 24 (1890), 332–338. Here, Heawood also investigated the coloring of maps on orientable surfaces other than the plane or sphere (see Chapter 3).

2

Coloring Maps and Graphs

As we saw in Chapter 1, the latter half of the 19th century witnessed much activity on the four-color problem of determining whether the countries of every map can be colored with just four colors in such a way that neighboring countries are colored differently. This problem's formulation by Francis Guthrie in 1852, its propagation by Augustus De Morgan, and its revival by Arthur Cayley, led to Alfred Kempe's attempted proof in 1879 and to its subsequent refutation in 1890 by Percy Heawood. Thus, by the end of the 19th century, the four-color problem had again become an open problem, one that would achieve great popularity and notoriety in the ensuing years.

We begin this chapter by describing various attempts to solve it, from the early 20th century to its eventual computer-assisted solution in 1976. Work on this problem led to related investigations into colorings of the edges or vertices of a map or graph, and specifically into coloring the vertices of a planar graph. These explorations then took on lives of their own, as we see in the second half of this chapter.

2.1 Coloring maps

Although Kempe's attempted proof of the *four-color theorem*, that the countries of every map can indeed be colored with four colors, was erroneous, it contained two ideas that would prove of crucial importance in subsequent attempts on the problem—the concepts of an *unavoidable set of configurations* and a *reducible configuration*, where a *configuration* is a collection of countries that is surrounded by a ring of other countries. Throughout this section we will mainly restrict our attention to *cubic maps*, where exactly three countries and three boundary edges meet at each meeting point of countries.

Using ideas derived from Euler's polyhedron formula, Kempe proved that every map must contain at least one country bounded by five edges or fewer—that is, the collection of four configurations (digon, triangle, quadrilateral, and pentagon) in Figure 2.1 is *unavoidable*, in the sense that every cubic map must contain at least one of these.

Kempe also attempted to prove that each of these configurations is *reducible*, in the sense that if a map contains such a configuration, then any coloring of the rest of this map with four colors can be extended to the configuration, either directly or after

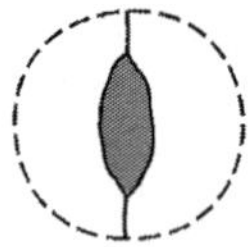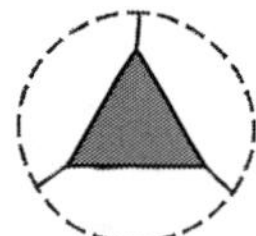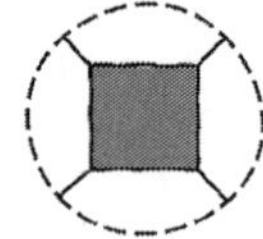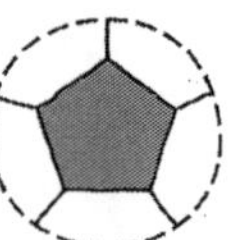

Figure 2.1. Kempe's unavoidable set of configurations

a simple modification—this provides the basis for a proof by induction. A reducible configuration cannot appear in a smallest counter-example to the four-color theorem.

Kempe observed that any country with up to three boundary edges is reducible, and he developed an argument to prove that a quadrilateral is reducible. Now known as a *Kempe-chain argument*, this involved the interchange of colors in two-colored portions of the map. However, his attempt to extend the idea to a pentagonal country was flawed, as was pointed out eleven years later by Heawood.[1]

Kempe also introduced the "dual problem" of coloring the vertices of a planar graph. As he remarked, if we lay a piece of tracing paper over a map, mark a point within each country (such as its capital city), and then join pairs of these points in countries that have a common boundary, we obtain a "linkage" (shown in Figure 2.2 with dashed edges). The problem is thereby transformed into that of coloring the points of the linkage with four colors in such a way that no two joined points receive the same color. As Kempe remarked: "What are the linkages which can be similarly lettered with no less than n letters? The classification of linkages according to the value of n is one of considerable importance." Kempe was thereby the first to define the chromatic number of graphs in general, not only planar ones.

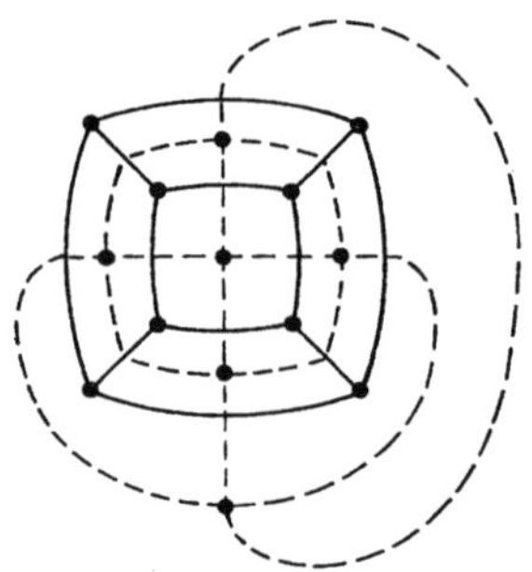

Figure 2.2. From maps to linkages

2.1.1 Two approaches. Once the error in Kempe's proof had been discovered, attempts to patch up the argument headed in two different directions—a search for new unavoidable sets that avoided the troublesome pentagon, and the hunt for further reducible configurations to supplement the digon, triangle, and quadrilateral.

Because the pentagon could not be proved reducible, attempts were made to replace it by other arrangements of countries. One author who did so was Paul Wernicke, a professor of modern languages at the State College of Kentucky with an interest in mathematics. Born in Berlin in 1866, he migrated to the United States in 1893 and became a naturalized American citizen. Returning to Germany for his graduate studies, he received a doctorate from the University of Göttingen in 1903 for a thesis on the coloring of maps, in which he proved that every map containing no country with fewer than five boundary edges must include either two adjacent pentagons or a pentagon adjacent to a hexagon;[2] this gives the unavoidable set in Figure 2.3. Wernicke's hope was that these two configurations would prove less troublesome than a single pentagon had been.

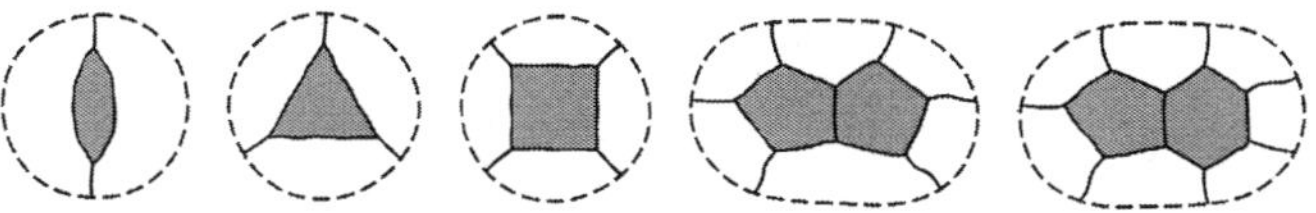

Figure 2.3. Wernicke's unavoidable set of configurations

Another Göttingen mathematician who failed to prove the four-color theorem was Wernicke's doctoral supervisor, Hermann Minkowski. Announcing the problem during his lectures on analysis situs (topology) in 1904–05, he informed his class that he could surely solve it. Over several weeks he attempted to construct a proof, but without success. Finally, to the accompaniment of a loud crash of thunder, he declared to his class that Heaven was angered by his arrogance, and that his attempts on the four-color theorem were also defective. Among the impressive list of mathematicians who attended these lectures were Max Born, Constantin Carathéodory, Dénes König, and Hermann Weyl.[3]

In 1912, two new American figures entered the scene. One of these was Oswald Veblen of Princeton University, who reformulated the four-color problem in the language of linear algebra, as we shall see in Chapter 4. He later presented lectures on analysis situs in the American Mathematical Society's Colloquium Series, and wrote a monograph on the subject that became a standard work of reference.[4]

The other was the Harvard mathematician George D. Birkhoff, who had a lifelong fascination with the four-color problem. In 1912, he took a more quantitative approach by introducing the concept of the *chromatic polynomial* of a map.[5] This polynomial gives the number of possible colorings of the map with a given number of colors; for example, if k colors are available, then the map in Figure 2.4 can be colored in $k(k-1)(k-2)^2$ different ways (not coloring the outer region). In his paper Birkhoff found a general expression for the coefficients of this polynomial.

By studying general properties of these polynomials, Birkhoff hoped to prove that if $P(k)$ denotes the number of ways of coloring a map with k colors, then $P(4) > 0$

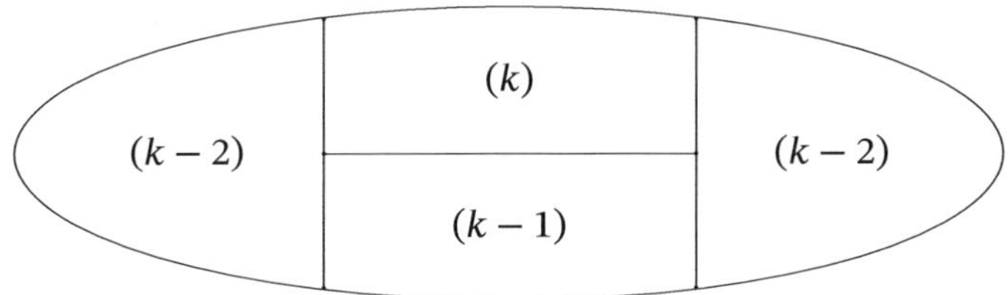

Figure 2.4. The chromatic polynomial of a map

for every map—that is, all maps can be colored with four colors. He continued to produce further papers on this topic, culminating in a lengthy posthumous survey with his Harvard University doctoral student Daniel C. Lewis; for many years their joint paper became the standard introduction to the subject.[6]

In a second classic paper, published in 1913, Birkhoff turned his attention to the study of reducible configurations, systematically investigating the colorings of rings of countries that surround a given configuration of countries.[7] After dealing with surrounding rings of three, four, and five surrounding countries and relating these to the four-color problem, he turned his attention to coloring rings of six countries. This more difficult situation led him, in particular, to consider the case of a ring of six countries surrounding four pentagons, a configuration that later came to be known as the *Birkhoff diamond* (see Figure 2.5). By showing that all thirty-one of the essentially different colorings of the surrounding ring can be extended to colorings with four colors of the internal pentagons, either directly or after Kempe-chain interchanges of colors, he deduced that the diamond configuration is reducible.

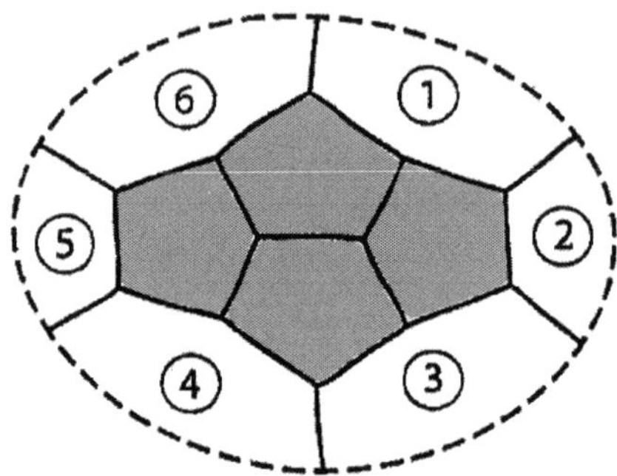

Figure 2.5. The Birkhoff diamond

In the 1920s, Birkhoff's pioneering results were developed by mathematicians in both Europe and America. In Belgium, Alfred Errera wrote a doctoral thesis on *The Coloring of Maps and Some Questions of Analysis Situs*, which featured another counterexample to Kempe's assertions about the pentagon.[8] He then developed his ideas on map coloring in several papers on the subject, and proved that a smallest counterexample to the four-color theorem cannot consist solely of pentagons and hexagons.

Meanwhile, in the United States, Philip Franklin's doctoral thesis for Princeton University, supervised by Veblen, was also on the four-color problem. Here, Franklin extended the results of Wernicke and Birkhoff by proving that every map that includes no digon, triangle, or quadrilateral must contain at least twelve pentagons.[9] Moreover, such a map must include a pentagon adjacent to two other pentagons, a pentagon adjacent to a pentagon and a hexagon, or a pentagon adjacent to two hexagons, thereby giving rise to a new unavoidable set (see Figure 2.6).

Using a counting argument, Franklin deduced that four colors suffice for all maps with at most twenty-five countries, so that any search for a counterexample was likely to prove difficult. Similar, but more complicated, arguments then led to this number being increased further:[10] to twenty-seven by Clarence N. Reynolds, Jr. of West Virginia in 1926–27, to thirty-one by Franklin in 1938, to thirty-five by Charles Winn from Egypt in 1940, to thirty-nine by Oystein Ore and Joel Stemple of Yale University in

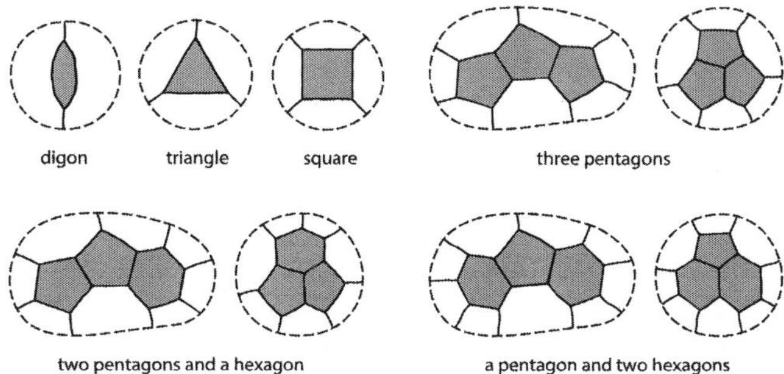

Figure 2.6. Franklin's unavoidable set of configurations

1968, and to ninety-five by Jean Mayer, Professor of French Literature at the University of Montpellier, around 1975. But there was still a long way to go ...

Until the 1930s, the only papers on unavoidable sets had been those of Kempe, Wernicke, and Franklin, but in 1940, in the last year of his life, the veteran French mathematician Henri Lebesgue, best known for his introduction of the Lebesgue integral in mathematical analysis, discovered some new unavoidable sets and presented them in a paper on some simple consequences of Euler's formula.[11]

2.1.2 Proving the four-color theorem. As the 20th century wore on, increasing numbers of reducible configurations of countries were being discovered—eventually, several thousands of them. Meanwhile, papers on unavoidable sets of configurations remained scarce.

It was not until the 1960s that books specially devoted to the four-color problem began to appear. The first of these, by Ore, was published in 1967, and two years later, the German mathematician Heinrich Heesch produced another book on the subject.[12] Heesch had been fascinated by the four-color problem since the 1930s as a student in Göttingen, and in a lecture at the University of Kiel in 1948 he explained to his audience (which included the young Wolfgang Haken) that the four-color problem would be solved if one could find *an unavoidable set of reducible configurations*—for if such an unavoidable set were to exist, then every map would have to include at least one of them, and yet each of them would be reducible and so could not appear in a smallest counterexample to the four-color theorem. No such counterexample could then exist, and the four-color theorem would be proved.

To investigate reducible configurations further, Heesch defined a configuration to be *D-reducible* if every coloring of the surrounding ring can be extended to the interior, either directly or after Kempe-chain interchanges of colors, and to be *C-reducible* if this can be carried out for a restricted set of colorings of the surrounding ring obtained by simplifying the configuration in an appropriate way. Heesch became convinced on probabilistic grounds that unavoidable sets of reducible configurations must exist, and he estimated that such a set could be found with not more than 9000 configurations in it.

Heesch also developed a way of testing unavoidability in cubic maps, later called (by Haken) *the method of discharging.* In order to investigate whether certain configurations formed an unavoidable set, Heesch assigned a "charge" of $6 - k$ to each

k-sided country, so that each pentagon received a unit charge, each hexagon received zero charge, and so on; it follows from Euler's polyhedron formula that the total charge over the whole map is 12. He next attempted to move the charges around the map so that no charge was created or destroyed. In many cases this could be done in such a way that a country with a positive charge after the change would provide one of the given configurations, confirming that the configurations did indeed form an unavoidable set.

Up to the 1970s, most investigators had attempted to prove the four-color theorem by constructing large collections of reducible configurations and trying to package them together into an unavoidable set. But the great breakthrough occurred when Haken, by then at the University of Illinois, teamed up with Kenneth Appel. Together, they adopted a different approach, preferring to develop increasingly large unavoidable sets of configurations and then to test these configurations for reducibility, modifying the unavoidable set as necessary whenever any irreducible configurations emerged. After working in this way for a few years, they eventually developed in 1976 an unavoidable set of nearly two thousand reducible configurations (later reduced to 1482). This proved the four-color theorem.

More precisely, working in Kempe's dual formulation of coloring the vertices of planar graphs (that is, the points of linkages), which had by then become the standard approach, Appel and Haken developed a range of discharging methods for producing unavoidable sets of configurations that had a good chance of being reducible; indeed, as they gained experience they gradually developed a "sixth sense" for spotting with high accuracy whether a given configuration of countries was likely to be reducible. These configurations were then tested for reducibility and replaced by new configurations when necessary. A few of their configurations (both D-reducible and C-reducible) appear in graph form in Figure 2.7, using Heesch's representation of each pentagon by a solid circle (•), each hexagon by a dot (·), and each octagon by a square (□).

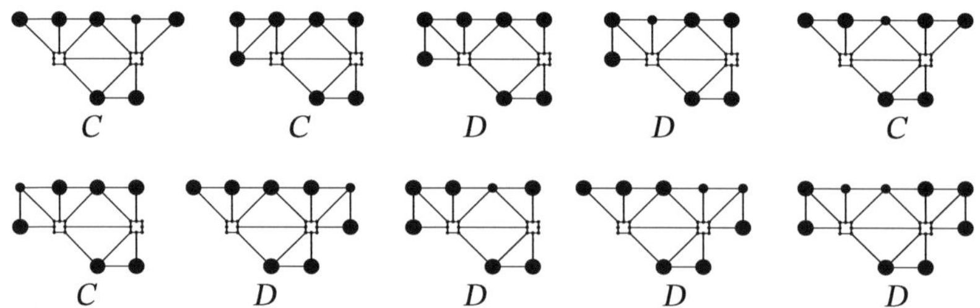

Figure 2.7. Some of Appel and Haken's configurations

Throughout their investigations, Appel and Haken made extensive use of computers, both to help them develop algorithms for producing unavoidable sets, and to test the resulting configurations for reducibility. By this time, the process had become extremely complicated: whereas the Birkhoff diamond had involved a surrounding ring of just six countries, Appel and Haken needed to investigate large numbers of configurations that involved rings of up to fourteen countries.

In 1976, they announced their success in a two-page research note for the American Mathematical Society, before presenting details of their proof in two lengthy papers in the *Illinois Journal of Mathematics*, one on unavoidable sets (which included a microfiche supplement), and the other, written with their graduate assistant John Koch, on reducible configurations. They also explained the ideas underlying their proof in popular articles in *New Scientist* and *Scientific American*.[13]

Perhaps inevitably, Appel and Haken's papers contained some errors, but these were soon corrected. Although some mathematicians were delighted to see the solution of such a long-standing unsolved problem, the achievement also caused much consternation around the mathematical world, because of its complexity and because it raised controversial philosophical questions about the validity of proofs that could not be checked by hand. Appel and Haken's computer programs were checked and rechecked, and various minor simplifications were suggested by a number of people, but no substantially new proof emerged. In 1989, with the collaboration of John Koch, they produced a lengthy book that contained further details of their proof and a printed version of their microfiche pages.[14]

In the 1990s, Neil Robertson and Paul Seymour, who for twenty years spent their summer months tackling difficult problems in graph theory (and frequently solving them), decided to join forces with two others, Daniel Sanders and Robin Thomas, in an attempt to prove the four-color theorem. Although they used essentially the same approach as Appel and Haken, they were more systematic in carrying it out, and their simplifications resulted in a more transparent proof based on an unavoidable set of 633 reducible configurations—less than half of those presented by Appel and Haken, but still a substantial number.[15] Shortly afterwards, Thomas presented an interesting survey of seemingly unrelated mathematical results (on the algebra of three-dimensional vectors, the divisibility of integers, and matrices and tensors) that turn out to be equivalent to the four-color theorem.[16]

By this time, computer-assisted solutions to mathematical problems had become more commonplace, and in 2004, the French computer scientist Georges Gonthier presented a fully machine-checked proof of the four-color theorem.[17] This was a formal language implementation and machine verification of the proof by Neil Robertson and his team. Doubters could at last consider the four-color problem to be solved.

Because D-reducible configurations are usually easier to deal with than the more general C-reducible ones, the question arose as to whether the four-color theorem could be proved by using only the former. In 2008, this question was answered in the affirmative by John Steinberger, who produced an unavoidable set of 2832 D-reducible configurations.[18] Whereas Appel and Haken's proof had involved configurations with surrounding rings of up to fourteen countries, Steinberger's rings extended to sixteen countries.

2.2 Coloring graphs

In 1880, the Scottish applied mathematician Peter Guthrie Tait attempted to simplify Kempe's ideas on the four-color problem by reformulating it in terms of coloring the boundary edges of a cubic map;[19] for example, given the map of Figure 2.8 whose countries have been assigned the four colors $A, B, C,$ and D, we can color their boundaries with 1, 2, and 3, in such a way that the edges at each boundary point use all three colors. To do so, we use use color 1 for edges bounding those countries colored $A\&B$ or $C\&D$, color 2 for edges bounding those countries colored $A\&C$ or $B\&D$, and color 3 on edges bounding those countries colored $A\&D$ or $B\&C$, as illustrated.

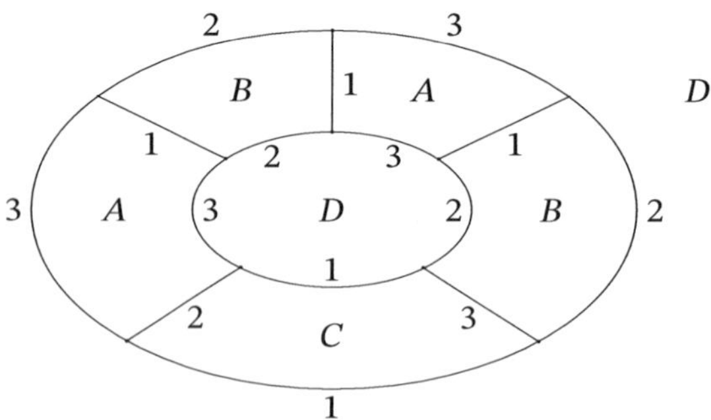

Figure 2.8. Coloring the edges of a map

Tait believed that a simple induction argument would prove that the edges of every cubic map can be so colored with three colors, and that, by reversing the process, he could deduce the four-color theorem for maps. But he was wrong—this edge-coloring result is just as difficult to prove as the original theorem.

2.2.1 Edge-colorings. Building on Tait's ideas, we can ask how many colors are needed to color the edges of an arbitrary given graph G so that any two adjacent edges (edges that meet at a common vertex) are colored differently. It is clear that if G has a vertex of degree k then k colors are required to color these edges, and it follows that if Δ is the largest vertex-degree in G, then at least Δ colors are needed to color the edges of G. But for some graphs, such as odd cycles or complete graphs with an odd number of vertices, more than Δ colors are required (see Figure 2.9).

Recall that a graph is *bipartite* if its set of vertices can be split into two parts so that every edge joins a vertex in one part to a vertex in the other. In 1916, in a fundamental paper relating to set theory and determinants,[20] the Hungarian mathematician Dénes König investigated such graphs, and showed that a graph is bipartite if and only if every cycle has an even number of edges. He also proved that, for bipartite graphs, the number of colors needed for its edges is always equal to the largest vertex-degree; for example, Figure 2.10 shows a bipartite graph with largest vertex-degree 4 and an edge-coloring with four colors.

For several years, little attention was paid to edge-colorings of graphs, but in 1949 Claude Shannon, the "father of information theory", investigated the coloring of wires in an electrical unit in which several wires may join the same pairs of terminals.[21] In

Figure 2.9. Coloring the edges of the cycle C_5 and the complete graph K_5.

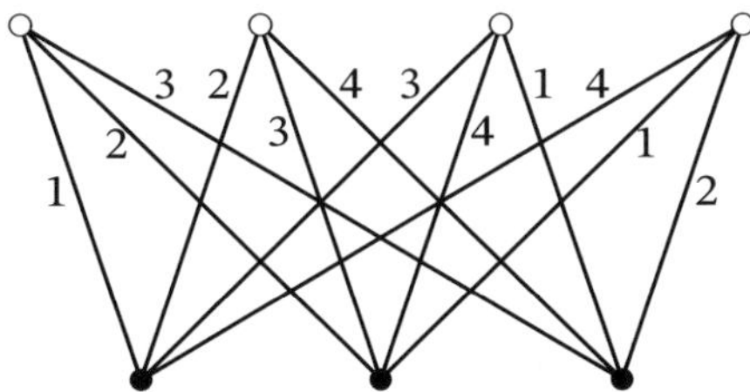

Figure 2.10. Coloring the edges of the complete bipartite graph $K_{3,4}$

these circumstances, where two vertices may be joined by more than one edge, we have a *multigraph*, and Shannon proved that if such a multigraph has largest vertex-degree Δ, then the number of colors needed to color its edges does not exceed $\frac{3}{2}\Delta$. The multigraph with three vertices joined in pairs by the same number d of edges shows that this cannot be improved because each vertex has degree $2d$ and all $3d$ edges must have different colors (see Figure 2.11).

In 1964, Shannon's result was extended in a classic paper by the Russian mathematician Vadim Vizing,[22] who proved that if at most p multiple edges join any pair of vertices in a multigraph G with maximum vertex-degree Δ, then the edges of G can always be colored with at most $\Delta + p$ colors. In particular, if G is a simple graph, then Vizing's theorem tells us that its edges can be colored with either Δ or $\Delta + 1$ colors—a spectacularly sharp result.

A year later, Vizing followed this paper with two others, in which he obtained several more edge-coloring results that are now regarded as fundamental.[23] A particularly useful result, sometimes called *Vizing's adjacency lemma*, tells us that, for connected simple graphs with three or more vertices, there are at least three vertices of maximum degree, and that such vertices appear throughout the graph. Since the appearance of Vizing's papers, considerable effort has been expended in trying to determine which simple graphs can be edge-colored with Δ colors and which ones require $\Delta + 1$ colors.

There has also been much recent work on edge-colorings of nonsimple multigraphs. In particular, Mark Goldberg, Ram Prakash Gupta, and Paul Seymour have independently conjectured that if the smallest number of colors needed to color the

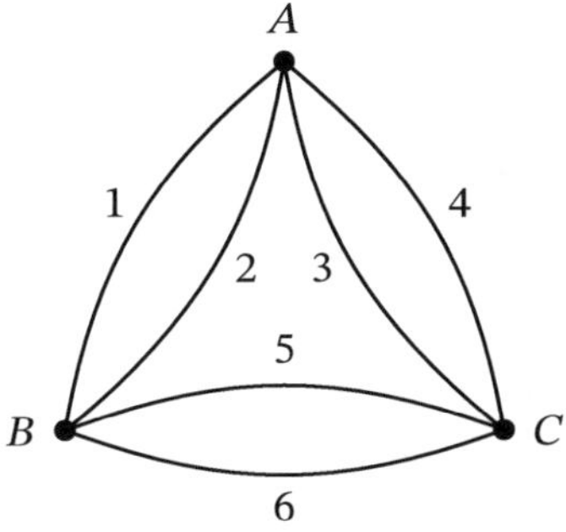

Figure 2.11. Coloring the wires of an electrical network

edges of a multigraph exceeds $\Delta + 1$, then it must equal

$$\max_{H} \left\lceil \frac{2m(H)}{n(H) - 1} \right\rceil,$$

where the maximum is taken over all subgraphs H of the graph with $m(H)$ edges and an odd number $n(H)$ of vertices;[24] because each color appears on at most $\frac{1}{2}(n(H)-1)$ edges of H, this quantity is a lower bound for the number of colors required. A proof of this conjecture provides a minimax theorem for edge-colorings, with possible algorithmic consequences.[25]

2.2.2 Vertex-colorings. We have seen how Kempe reformulated the four-color problem in terms of coloring the points of a linkage or graph. In this formulation, the problem is to decide whether the vertices of every graph drawn in the plane can be colored with four colors so that any two vertices joined by an edge are colored differently. More generally, we can ask how many colors are required for the vertices of an arbitrary given graph G. This number is the *chromatic number* $\chi(G)$ of G; for example, if K_n is the *complete graph* on n vertices then $\chi(K_n) = n$, and if G is the graph of Figure 2.12, then $\chi(G) = 3$.

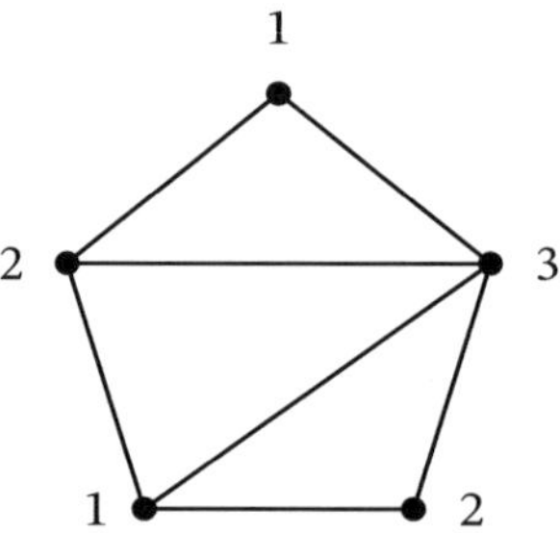

Figure 2.12. Coloring the vertices of a graph

It is not difficult to prove that if G is a graph in which the largest vertex-degree is Δ, then its vertices can be colored with at most $\Delta + 1$ colors. In a classic paper of 1941, the English mathematician Leonard Brooks showed that for a connected graph this upper bound can be reduced to Δ, except when G is a complete graph or an odd cycle.[26]

A different approach to vertex-coloring was taken in 1943 by the Swiss mathematician Hugo Hadwiger, who conjectured that every connected graph with chromatic number n can be "contracted" to the complete graph with n vertices.[27] Here, we *contract* an edge $e = vw$ by replacing v and w by a single vertex x adjacent to those vertices that are adjacent to v or w or both (see Figure 2.13), and a graph G is *contractible* to a graph H if H can be obtained from G by a succession of edge-contractions. An example of contracting four independent edges is shown in Figure 2.14.

Hadwiger proved his conjecture for $n = 2, 3$, and 4. In 1937, the German mathematician Klaus Wagner proved that the truth of this conjecture for $n = 5$ is equivalent to the four-color theorem, then still unproved. Since then, Hadwiger's conjecture has been proved for $n = 6$ by Robertson, Seymour, and Thomas, who showed that its truth is also equivalent to the four-color theorem, but it remains unproved in general.[28] We return to this topic in Chapter 6.

Figure 2.13. Contraction of an edge

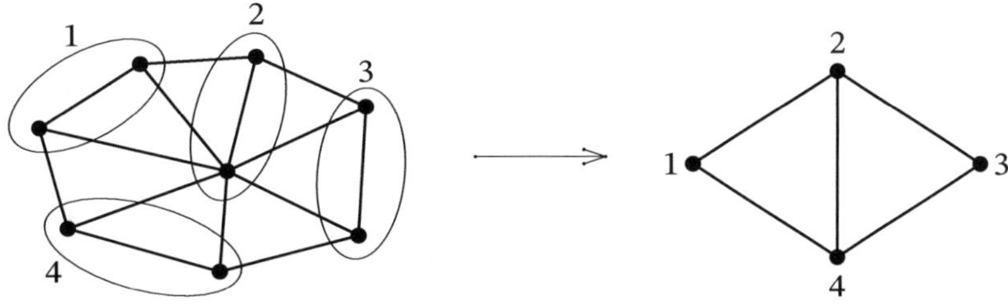

Figure 2.14. Contraction of a graph

In the 1950s and 1960s, several people investigated the chromatic numbers of graphs, and much progress was made. In particular, Gabriel Andrew Dirac in England and Tibor Gallai (pupil of König in Hungary) both investigated the properties of *critical graphs*; these are graphs for which the removal of any vertex or edge lowers the chromatic number, and because they tend to have more structure than graphs in general their study has proved fruitful.[29] Dirac had moved from Hungary to England in 1937 at age 12, where his mother had married the Nobel laureate physicist, P. A. M. Dirac. After the war he studied in London, where he was introduced to graph theory, and this led to Dirac's doctoral thesis in 1951 where critical graphs were first defined. Gallai and Dirac eventually met in 1959 at the first ever conference in graph theory, at Dobogókő in Hungary.

Yet another investigation has been into graphs that contain few triangles (cycles with three edges). In 1958, the German mathematician Herbert Grötzsch proved that the vertices of every planar graph with at most three triangles can be colored with three colors.[30]

More recently, attention has shifted also to other types of coloring numbers. One of these is *total coloring*, introduced in 1965 in the doctoral thesis of Mehdi Behzad and in seminars by Vizing given in Novosibirsk around the same time.[31] Here, both the vertices and the edges of a graph are assigned colors in such a way that adjacent vertices have different colors, adjacent edges have different colors, and each edge is colored differently from its incident vertices. The *total coloring conjecture*, which remains unproved, is that any simple graph with maximum degree Δ has a total coloring with at most $\Delta + 2$ colors.

Another much-studied type of coloring was introduced by Vizing, and independently by Paul Erdős, Arthur L. Rubin, and Herbert Taylor.[32] A *list coloring* of a graph

G is a coloring in which each vertex is first assigned a list of colors after which each vertex is properly assigned a color from its list (with adjacent vertices still having different colors). An example of their use appears in an elegant note by the Danish mathematician Carsten Thomassen, who gave a list-generalization of the five-color theorem for planar graphs, proving it by induction in a simple manner that, like Heawood's proof of the five-color theorem, avoided the recoloring of previously colored subgraphs.[33]

2.3 Conclusion

We conclude this chapter with a quotation by William T. Tutte, one of the most important combinatorial mathematicians of the 20th century. During World War II he worked at Bletchley Park and was instrumental in deciphering German secret codes. After the war he moved to Canada and worked for most of his later life at the University of Waterloo, where he founded the Department of Combinatorics and Optimization in 1967.

In 1976, when the four-color theorem was first proved, it was feared in some quarters that this was the end of the line and that little was left for graph-theorists to get their teeth into. In a delightful popular article, Tutte contradicted this idea, declaring the four-color problem as merely a beginning.[34] Citing Hadwiger's conjecture and other unsolved problems that include the four-color theorem as a special case, he observed that

> *The Four Colour Theorem*
> *is the tip of the iceberg,*
> *the thin end of the wedge*
> *and the first cuckoo of Spring.*

Further Reading

For further information on the topics of this chapter, see the following multi-authored volume:

> Lowell W. Beineke and Robin J. Wilson (eds.), *Topics in Chromatic Graph Theory*, Cambridge Univ. Press (2015).

> See, in particular, the survey chapters on "Chromatic polynomials" by Bill Jackson, "Brooks's theorem" by Michael Stiebitz and Bjarne Toft, "Edge-colourings" by Jessica McDonald, and "List colourings" by Michael Stiebitz and Margit Voigt.

Other books on graph coloring in general are

> Tommy R. Jensen and Bjarne Toft, *Graph Coloring Problems*, Wiley-Interscience (1995);

> Gary Chartrand and Ping Zhang, *Chromatic Graph Theory*, CRC Press (2009);

> M. Stiebitz, T. Schweser, and B. Toft, *Brooks' Theorem – Graph Coloring and Critical Graphs*, Springer (2024).

For edge-colorings in particular, see the following:

S. Fiorini and R. J. Wilson, *Edge-Colourings of Graphs*, Pitman (1977);

Michael Stiebitz, Diego Scheide, Bjarne Toft, and Lene M. Favrholt, *Graph Edge Coloring: Vizing's Theorem and Goldberg's Conjecture*, Wiley (2012).

For further information on the history and proof of the four-color theorem, see Robin Wilson, *Four Colors Suffice* (revised color edition), Princeton Univ. Press (2014).

Some earlier books on the four-color problem include

Oystein Ore, *The Four-Color Problem*, Academic Press (1967);

Thomas L. Saaty and Paul C. Kainen, *The Four-Color Problem: Assaults and Conquests*, McGraw-Hill (1977);

David Barnette, *Map Coloring, Polyhedra and the Four-Color Problem*, Dolciani Math. Expositions 8, MAA (1983);

Rudolf Fritsch and Gerda Fritsch, *The Four-Color Theorem: History, Topological Foundations, and Ideas of Proof*, Springer (1999).

Notes and References

[1]The references for Kempe's and Heawood's papers appear in Chapter 1. Heawood subsequently published several further papers on the four-color problem, the last of these appearing in his 90th year.

[2]P. Wernicke, Über den kartographischen Vierfarbensatz, *Math. Ann.* 58 (1904), 413–426. Wernicke subsequently presented an attempted proof of the four-color theorem to the American Mathematical Society, but this has not survived.

[3]This story appears in Constance Reid, *Hilbert*, Springer (1970), 92–93. In König's notes from these lectures of Minkowski, the sentence stating that the four-color problem theorem would be proved is crossed out. A facsimile of parts of these notes is contained in the 1986 Teubner reprint of König's monograph *Theorie der endlichen und unendlichen Graphen*, Akademische Verlagsgesellschaft, Leipzig (1936).

[4]For Veblen's reformulation of the four-color problem, see O. Veblen, An application of modular equations in analysis situs, *Ann. of Math.* (2) 14 (1912/13), 86–94, and Chapter 4 of this book. His Colloquium lectures were published as *Analysis Situs*, Amer. Math. Soc. Colloq. Publ. Vol. 5 (1922).

[5]George D. Birkhoff, A determinant formula for the number of ways of coloring a map, *Ann. of Math.* (2) 14 (1912–13), 42–46.

[6]Birkhoff's later papers on chromatic polynomials appeared in 1930 and 1934, and the posthumous survey (written mainly by Lewis) was G. D. Birkhoff and D. C. Lewis, Chromatic polynomials, *Trans. Amer. Math. Soc.* 60 (1946), 355–451.

[7]George D. Birkhoff, The reducibility of maps, *Amer. J. Math.* 35 (1913), 115–128.

[8]Alfred Errera's thesis, *Du coloriage des cartes et de quelques questions d'Analysis Situs*, was published by Falk Fils and Gauthier-Villars in 1921.

[9]Philip Franklin, The four color problem, *Amer. J. Math.* 44 (1922), 225–236.

[10]The results listed here can be found in the following papers:

• C. N. Reynolds Jr., On the problem of coloring maps in four colors, I, II, *Ann. of Math.* (2) 28 (1926–27), 1–15 and 477–492;

• Philip Franklin, Note on the four color problem, *J. Math. Phys.* 16 (1938), 172–184;

• C. E. Winn, On the minimum number of polygons in an irreducible map, *Amer. J. Math.* 62 (1940), 406–416;

- O. Ore and J. Stemple, Numerical calculations on the four-color problem, *J. Combin. Theory* 8 (1970), 65–78.

Jean Mayer's result was unpublished.

[11] Henri Lebesgue, Quelques consequences simples de la formula d'Euler, *J. Math. Pures et Appl.* 9 (1940), 27–43.

[12] These two books are

- O. Ore, *The Four-Color Problem*, Academic Press (1967);

- H. Heesch, *Untersuchungen zum Vierfarbenproblem*, B. I. Hochschuleskripten, 810/810a/810b, Bibliograph-isches Institut, Mannheim-Wien-Zürich, 1969.

[13] The publications referred to in this paragraph are

- K. Appel and W. Haken, Every planar map is four colorable, *Bull. Amer. Math. Soc.* 82 (1976), 711–712;

- K. Appel and W. Haken, Every planar map is four colorable, Part I: Discharging, *Illinois J. Math.* 21 (1977), 429–490;

- K. Appel, W. Haken, and J. Koch, Every planar map is four colorable, Part II: Reducibility, *Illinois J. Math.* 21 (1977), 491–567;

- K. Appel, The proof of the four-colour theorem, *New Scientist* 72 (21 October 1976), 154–155;

- K. Appel and W. Haken, The solution of the four-color-map problem, *Scientific American* 237 (4) (October 1977), 108–121.

[14] Kenneth Appel and Wolfgang Haken, *Every planar map is four colorable*, Amer. Math. Soc., 1989.

[15] For a detailed account of their approach, see N. Robertson, D. Sanders, P. Seymour, and R. Thomas, The four-colour theorem, *J. Combin. Theory (B)* 70 (1997), 2–44.

[16] Robin Thomas, An update on the four-color theorem, *Notices Amer. Math. Soc.* 45 (August 1998), 848–859.

[17] Georges Gonthier, Formal proof—the four-color theorem, *Notices Amer. Math. Soc.* 55 (December 2008), 1382–1393.

[18] John P. Steinberger, An unavoidable set of D-reducible configurations, *Trans. Amer. Math. Soc.* 362 (2010), 6633–6661.

[19] P. G. Tait, On the colouring of maps, *Proc. Roy. Soc. Edinburgh* 10 (1878–80), 501–503. Around this time, Tait produced several incorrect proofs of the four-color problem, but he is credited with his equivalent reformulation of the problem in terms of edge-colorings.

[20] D. König, Über Graphen und ihre Anwendung auf Determinanten-theorie und Mengen-lehre, *Math. Ann.* 77 (1916), 453–465.

[21] C. E. Shannon, A theorem on coloring the lines of a network, *J. Math. Phys.* 28 (1949), 148–151.

[22] V. G. Vizing, On an estimate of the chromatic class of a p-graph [in Russian], *Diskret. Analiz* 3 (1964), 25–30; an English translation appears in Stiebitz *et al.* (see above).

[23] V. G. Vizing, The chromatic class of a multigraph [in Russian], *Kibernetika* 1 (3) (1965), 29–39; English translation: *Cybernet. Systems Anal.* 1 (1965), 32–41. See also V. G. Vizing, Critical graphs with a given chromatic number [in Russian], *Discret. Analiz* 5 (1965), 9–17; an English translation appears in Stiebitz *et al.* (see above).

[24] This conjecture appears in

- M. K. Goldberg, On multigraphs of almost maximal chromatic class [in Russian], *Diskret. Analiz* 23 (1973), 3–7;

- R. P. Gupta, On the chromatic index and the cover index of a multigraph, *Theory and Applications*, Lecture Notes in Math. 642, Springer (1978), 91–110;

- P. Seymour, On multicolourings of cubic graphs, and conjectures of Fulkerson and Tutte, *Proc. London Math. Soc.* 38 (1979), 423–460.

[25] A proof of this conjecture was announced in 2019 by Guantao Chen, Guanging Jing, and Wenan Zang (arXiv:1901.10316). Earlier, in 2007, a special case had been obtained by Diego Scheide, using ideas of the Russian Tashkīnov that generalized Vizing's approach.

[26]R. L. Brooks (1941), On colouring the nodes of a network, *Proc. Cambridge Philos. Soc.* 37 (1941), 194–197.

[27]H. Hadwiger, Über eine Klassification der Streckenkomplexe, *Vierteljschr. Naturforsch. Ges. Zürich* 88 (1943), 133–142.

[28]The publications referred to here are K. Wagner, Über eine Eigenschaft der ebenen Komplexe, *Math. Annalen* 114 (1937), 570–590; and N. Robertson, P. D. Seymour, and R. Thomas, Hadwiger's conjecture for K_6-free graph, *Combinatorica* 13 (1993), 279–361.

[29]The papers on critical graphs by Dirac and Gallai are

• G. A. Dirac, A property of 4-chromatic graphs and some remarks on critical graphs, *J. London Math. Soc.* 27 (1952), 85–92;

• T. Gallai, Critical graphs, *Theory of Graphs and its Applications*, Academic Press (1964), 43–45;

Gallai's paper provides surveys of two earlier lengthy papers of his, published in German in 1963.

Critical graphs feature in a far-reaching generalization of Brooks's theorem, obtained in 2014, when the minimum number of edges in a critical 4-chromatic graph with n vertices was shown to be $\lfloor 5n/3 \rfloor$. This number is also known exactly when the chromatic number $k > 4$ and $n \equiv 1 \pmod{k-1}$; see A. V. Kostochka and M. Yancey, Ore's conjecture for $k = 4$ and Grötzsch's theorem, *Combinatorica* 34 (2014), 323–329.

[30]H. Grötzsch, Ein Dreifarbensatz für dreikreisfreie Netze auf der Kugel, *Wiss. Z. Martin-Luther-Univ., Halle–Wittenberg. Math.–Natur. Reihe* 8 (1959), 109–120.

[31]M. Behzad, *Graphs and their Chromatic Numbers*, Ph.D. thesis, Michigan State University, 1965; and V. G. Vizing, Some unsolved problems in graph theory [in Russian], *Uspekhi Mat. Nauk* 23 (1968), 117–134; an English translation appears in *Russian Math. Surveys* 23 (1968), 125–141.

[32]V. G. Vizing, Vertex colorings with given colors [in Russian], *Diskret. Analiz* 29 (1976), 3–10; and P. Erdős, A. L. Rubin, and H. Taylor, Choosability in graphs, *Congr. Numer.* 26 (1979), 125–127.

[33]Carsten Thomassen, Every planar graph is 5-choosable, *J. Combin. Theory* (B) 62 (1994), 180–181.

[34]W. T. Tutte, Colouring problems, *Math. Intelligencer* 1 (1978), 72–75.

3

Graphs on Surfaces

The early history of planar graphs was intimately tied to Euler's polyhedron formula, sometimes called the "fundamental theorem of topological graph theory" (see Chapter 1). It follows from this result that if a simple connected graph with $n(\geq 3)$ vertices can be drawn on a plane or sphere with no edges crossing, then it cannot have more than $3n - 6$ edges, and from this we can deduce that such a graph must have a vertex of degree 5 or less, a fact that has been used many times in studies on the coloring of maps and graphs. These results also arise in Section 3.1, where we investigate those graphs that can be drawn without crossings on the plane or sphere. Here, the most important result is a celebrated theorem of the Polish mathematician Kazimierz Kuratowski, which tells us which graphs allow such a drawing. We also consider some alternative characterizations of such graphs. Throughout this chapter, unless it is specified otherwise, the graphs being considered are simple.

In Section 3.2 we investigate the drawing of graphs in the plane with the fewest crossings possible. Our initial focus is on complete graphs and complete bipartite graphs. However, their crossings are known precisely only in limited cases, so a more extensive family involving cycles is given. The section concludes with families of graphs in which there are many intersections of given types.

Section 3.3 concerns the drawing of graphs on surfaces other than a sphere. Beginning with the generalization of Euler's formula to general surfaces, this chapter culminates in a spectacular result of Robertson and Seymour on graph minors (to be discussed further in Chapter 8). Another major achievement is a result of Gerhard Ringel and J. W. T. Youngs which tells us on which surfaces a given complete graph can be drawn, but no good method is known that tells us in general on which surfaces a given graph can be drawn—indeed, this is a difficult problem, even for cubic graphs.

3.1 Graphs in the plane

The most famous result on graphs in the plane was discovered, not directly from the coloring of maps, but by looking first at polyhedra. A drawing of a connected graph G in the plane places the vertices of G at points in the plane, with adjacent vertices joined by simple curves; here, a "simple curve" is a continuous one-to-one mapping from the unit interval $[0, 1]$ to the plane. A drawing of G in which no edges cross is an *embedding*, and such an embedding is *cellular* if every region is connected and has no holes.

If G can be embedded in the plane, then it is *planar* and such an embedding is sometimes called a *plane drawing* of G (or simply a *plane graph*). Any such drawing divides those points of the plane that do not lie on an edge into *regions* (or *faces*) and, for a connected graph (not necessarily simple), the number of these regions is given by the analogue of Euler's polyhedron formula:

$$\text{(number of regions)} = \text{(number of edges)} - \text{(number of vertices)} + 2.$$

3.1.1 Characterizing planar graphs. A natural question asks which graphs are planar, and this was answered by Kuratowski,[1] who first presented it at a meeting of the Polish Mathematical Society in Warsaw in 1929, before publishing it in the following year. As noted above, Euler's formula implies that a simple connected planar graph with $n(\geq 3)$ vertices cannot have more than $3n - 6$ edges, and therefore the complete graph K_5 (with its 5 vertices and 10 edges) is nonplanar. Furthermore, a simple bipartite planar graph with $n(\geq 3)$ vertices cannot have more than $2n - 4$ edges, and so the complete bipartite graph $K_{3,3}$ (with its 6 vertices and 9 edges) is also non-planar (see Figure 3.1). Kuratowski's elegant result asserts that these two graphs are essentially the only obstacles to planarity.[2]

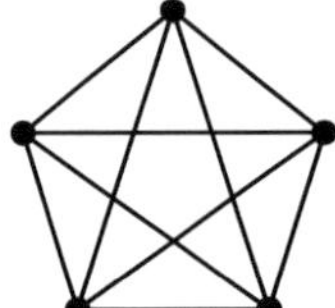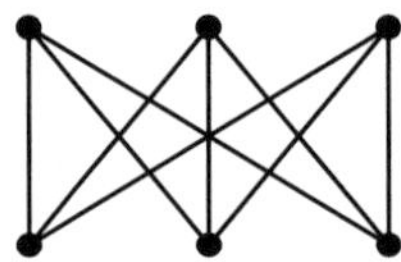

Figure 3.1. The two basic nonplanar graphs, K_5 and $K_{3,3}$

The word "essentially" here means that every nonplanar graph has a subgraph that resembles one of these. More precisely, a *subdivision* of an edge in a graph replaces that edge with a path of length 2, thereby adding a new vertex within the edge, and two graphs are called *homeomorphic* if each can be obtained from the same graph by such edge-subdivisions. Figure 3.2 shows a graph that is homeomorphic to $K_{3,3}$.

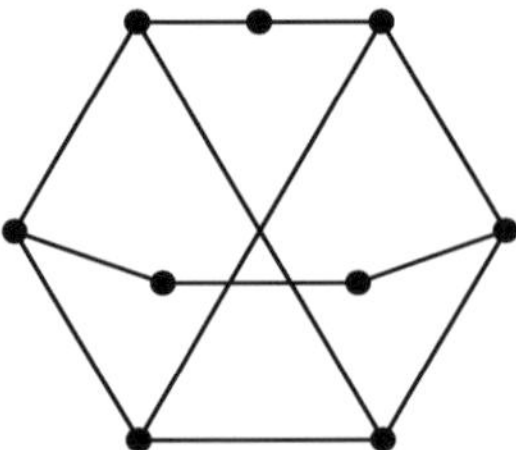

Figure 3.2. A graph homeomorphic to $K_{3,3}$

Kuratowski's theorem can then be stated as follows.

Theorem 3.1 (Kuratowski's theorem). *A graph is planar if and only if it has no subgraph homeomorphic to K_5 or $K_{3,3}$.*

Around the same time, Karl Menger[3] had proved a more restricted version, that every nonplanar cubic graph has a subgraph that is homeomorphic to $K_{3,3}$. Also in the same year that Kuratowski's paper appeared, Orrin Frink and P. A. Smith announced a proof of the same result,[4] but because of Kuratowski's priority they withdrew their paper and did not receive the credit they deserved. Since then, several other characterizations of planar graphs have been discovered, some of which we mention below.

Central to the study of the planarity of graphs is the concept of *duality*, which we have already met in connection with the linkages that Kempe derived from his maps (see Chapter 2). If G is a plane drawing of a connected planar graph, with multiple edges and loops allowed, a new vertex can be placed within each region and for each edge join the pair of new vertices on the two sides of the edge by a new edge. The graph thus formed by the new vertices and new edges is also a plane graph, called the *dual graph* of G and denoted G^*. If G has n vertices, m edges, and r regions, then G^* has r vertices, m edges, and n regions. For example, the dual of a cycle with m edges is a graph with two vertices joined by m edges. If an edge is a loop, then the corresponding dual edge is a *bridge* (defined as an edge not contained in any cycle), and vice versa. Note that $(G^*)^*$ is effectively the same graph as G (that is, the two graphs have exactly the same structure; this is expressed by saying that they are *isomorphic*, so that this is indeed a dual operation).

In the 1930s, graph theory became more structured and more "mathematical", in particular when the doctoral thesis of the American mathematician Hassler Whitney led to major contributions to the subject. In particular, in 1932 Whitney presented his results in an extended paper in which he investigated homeomorphic graphs, developed an abstract combinatorial theory of duality that does not rely on the above geometric construction, and proved that a graph is planar if and only if it has such an abstract dual.[5]

Whitney also showed that if a set of edges form a cycle in a connected plane graph, then the corresponding edges of a dual graph form a cut-set, and conversely. (Here, a *cut-set* is a minimal set of edges whose deletion splits the graph into two pieces.) Further contributions by Whitney included papers on the coloring of graphs, connectivity, and ways of classifying graphs.[6] In 1935, he showed how his graph-theoretical ideas have a natural setting in the more general idea of a *matroid*, which simultaneously generalizes notions of independence in both linear algebra and graph theory; for the latter, a set of edges of a graph is *independent* if it contains no cycles. Whitney's final contribution to graph theory was in 1972, when he and W. T. Tutte wrote an important paper on Kempe chains and the four-color problem.[7]

In 1937, Klaus Wagner found the following further characterization that closely resembles Kuratowski's theorem, but involves repeated contractions of edges rather than subdivisions.[8]

Theorem 3.2 (Wagner's theorem). *A graph is planar if and only if it has no subgraph that is contractible to K_5 or $K_{3,3}$.*

A version of this result also appears in a short paper of Frank Harary and W. T. Tutte,[9] that explains how the theorems of Kuratowski and Wagner are (in a certain sense) duals of one another—deleting an edge of a planar graph corresponds to contracting the corresponding edge in its dual, and vice versa.

Around this time Wagner wrote several papers on graph theory that included some results that were later credited to others. One of these states that if a simple graph can be drawn in the plane with no edges crossing, then this can be done in such a way that its edges are all straight-line segments, called a *linear plane graph* (see Figure 3.3). This result is generally known as Fáry's theorem, even though it had been proved over a decade earlier by Wagner.[10]

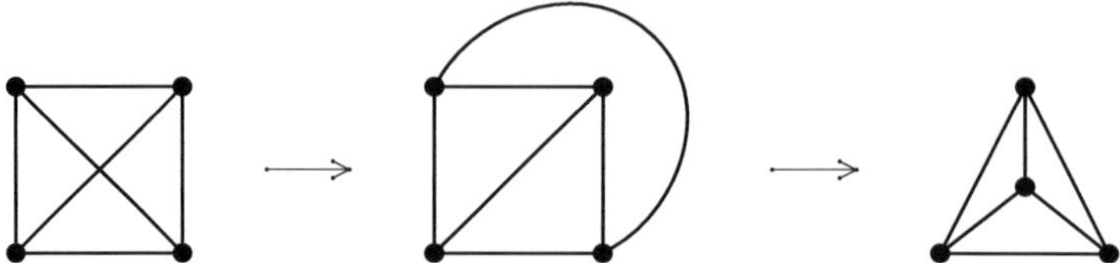

Figure 3.3. Drawing a planar graph with line segments for edges

Yet another elegant extension was given by Walter Schnyder,[11] who showed that every simple planar graph with n vertices has a linear embedding in which each vertex has integer coordinates within an $(n-1) \times (n-1)$ square. Figure 3.4 shows a planar graph with the five points $(1, 1)$, $(2, 2)$, $(3, 2)$, $(3, 4)$, and $(4, 1)$ as its vertices and the nine edges as its line segments.

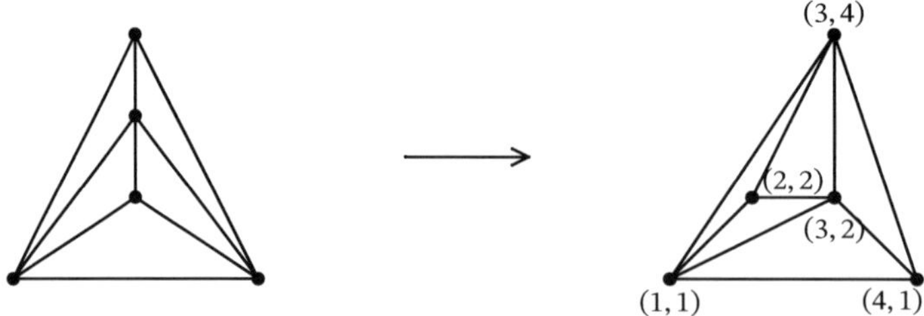

Figure 3.4. A linear plane graph with integer coordinates

Investigations into planar graphs have continued over the past fifty years and, in particular, algorithmic questions have been posed and solved. We discuss these in Chapter 8, where we consider questions of complexity.

3.1.2 Biplanarity and thickness.

Given two graphs with the same vertex-set V and with edge-sets E_1 and E_2, their *union*[12] is the graph with vertex-set V and edge-set $E_1 \cup E_2$. A graph is called *biplanar* if it is the union of two planar graphs. For example, the complete graph K_8 is not planar, but is biplanar because it is the union of the two graphs in Figure 3.5.

Which complete graphs are biplanar? By Euler's formula, a planar graph with n vertices has at most $3n - 6$ edges, and so a biplanar graph with 11 vertices can have

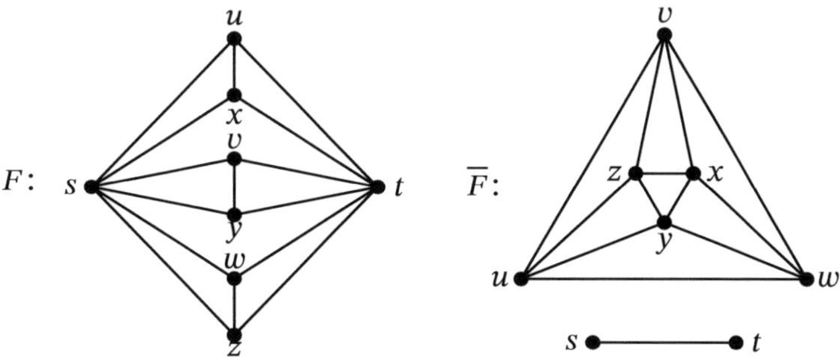

Figure 3.5. Two plane graphs with K_8 as their union

at most 54 edges. However, K_{11} has 55 edges, and hence is not biplanar, and neither is K_n for any larger n. It turns out that K_9 and K_{10} are also not biplanar, although K_9 becomes biplanar if a single edge is deleted.

Biplanar graphs have been studied in a variety of ways.[13] For instance, an interesting perspective of a planar graph is that its vertices can be placed in any specified position in a planar fashion. Hence, any biplanar graph G can have its vertices placed in identical positions on two plates of glass (or the two sides of a plate of glass) and its edges split between the two surfaces so that their union is G.

It is well known (see Chapter 8) that there are polynomial algorithms for determining whether a given graph is planar, but this is not the case for biplanarity. In fact, it was shown by Anthony Mansfield[14] that the problem of difficulty in determining biplanarity is included in an **NP**-completeness property that will be discussed in that chapter.

An extension of map-coloring in the plane concerns a planet with a moon, where each country on the planet can have one region on the moon (colored the same), and with neighboring regions colored differently as before.[15] How many colors are needed to color any planet-moon combination? In graphical terms this question, originally posed by Gerhard Ringel in 1959, asks for the maximum chromatic number of a biplanar graph.

This question is still unresolved and has a surprisingly wide range of possible answers. Just as the 6-color theorem for planar graphs follows from the existence of a vertex of degree 5 or less, so a similar argument yields a 12-color theorem for biplanar graphs. On the other hand, Thom Sulanke showed that the graph obtained by removing the edges of a 5-cycle from the complete graph K_{11} is biplanar with chromatic number 9. The maximum chromatic number of a biplanar graph is known only to lie in the range from 9 to 12 with several known 9-chromatic graphs.[16]

The question of whether the complete graph K_9 is biplanar was not easy to answer, but two proofs of its nonbiplanarity appeared around the same time, by Joseph Battle, Frank Harary, and Yukihiro Kodama in 1962, and by W. T. Tutte in 1963.[17] The result then inspired Tutte to define the *thickness* $\tau(G)$ of a graph G as the least number of planar subgraphs whose union is G.[18] Due to its significance in Very Large Scale Integration (VLSI) design, this is an important general concept, but here we discuss only complete graphs and complete bipartite graphs.

This eventually led to determining to the thickness of all complete graphs. The first substantial result was the thickness of five of every six complete graphs.[19]

Theorem 3.3 (Complete graph thickness theorem). *The thickness of the complete graph K_n is*

$$\tau(K_n) = \left\lceil \frac{n+7}{6} \right\rceil, \text{ except that } \tau(K_9) = \tau(K_{10}) = 3.$$

The first family of graphs for which an arbitrarily large thickness was determined was the complete bipartite graphs; there are no known exceptions and the unknown cases are quite rare.[20]

Theorem 3.4 (Complete bipartite graph thickness theorem). *The thickness of the complete bipartite graph $K_{r,s}$ is*

$$\tau(K_{r,s}) = \left\lceil \frac{rs}{2(r+s-2)} \right\rceil,$$

except possibly when r and s are both odd with $s \geq r+6$ and there is an integer k with $\frac{r+5}{4} \leq k \leq \frac{r-3}{2}$ for which

$$s = \left\lfloor \frac{2k(r-2)}{r-2k} \right\rfloor.$$

3.2 Crossing numbers

Another area of topological graph theory involves the drawing of arbitrary graphs in the plane, whether they are planar or not. For any such graph G, the minimum number of crossings in a plane drawing is its *crossing number* $\mathrm{cr}(G)$, so for every planar graph G, $\mathrm{cr}(G) = 0$. In this section we consider the crossing numbers of three families of graphs, and some variations on how one draws graphs.

3.2.1 Complete bipartite graphs.
Over the years many people have tried to draw figures in as simple a way as possible, but it was apparently not until World War II that the topic received any systematic mathematical treatment. Circumstances, involving the transportation of bricks in a labor camp by wagons on tracks from several ovens to several depots, led the Hungarian mathematician Paul Turán, imprisoned and working there, to seek the crossing number of the complete bipartite graph $K_{r,s}$ with r vertices in one set and s vertices in the other.[21] A full answer has proved elusive; indeed there are values r_0 and s_0 for which the crossing number of $K_{r,s}$ is not known if $r \geq r_0$ and $s \geq s_0$.

The first paper on this topic, claiming to answer Turán's brick-factory problem, was written by the Polish mathematician Kazimierz Zarankiewicz in 1954.[22] Unfortunately, his argument had an irremediable flaw, giving an upper bound for $\mathrm{cr}(K_{r,s})$ rather than an exact result. Although his upper bound

$$\mathrm{cr}(K_{r,s}) \leq \left\lfloor \frac{r}{2} \right\rfloor \left\lfloor \frac{r-1}{2} \right\rfloor \left\lfloor \frac{s}{2} \right\rfloor \left\lfloor \frac{s-1}{2} \right\rfloor$$

gives the correct value when r or s is at most 6, or when one is 7 or 8 and the other is 7, 8, 9, or 10, proving equality in the formula for $\mathrm{cr}(K_{r,s})$ for all higher values remains open. Representative of a general construction, an optimal drawing of $K_{5,7}$, giving Zarankiewicz's upper bound of 36, is shown in Figure 3.6.

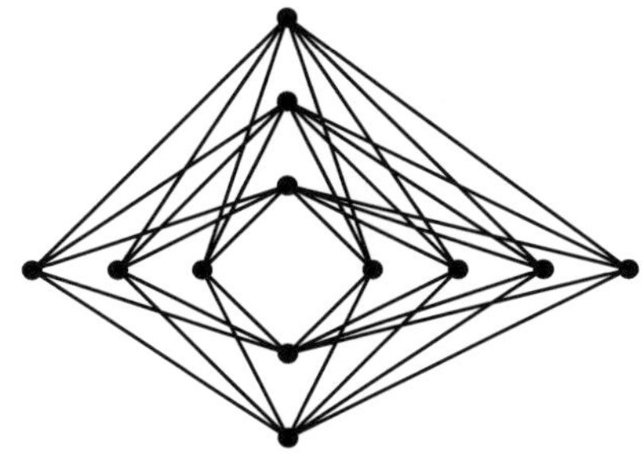

Figure 3.6. An optimal drawing of $K_{5,7}$

3.2.2 Complete graphs. Although it may seem a more natural problem, determining the crossing number of the complete graph K_n was apparently not investigated until some time later. An introduction to this problem was given by Richard Guy,[23] but earlier records of it can be traced back to some notes of the English artist and amateur mathematician Anthony Hill;[24] his construction placed the vertices in two concentric circles, with all pairs joined appropriately (see Figure 3.7 for the case $n = 8$), and yields the upper bound

$$\operatorname{cr}(K_n) \le \frac{1}{4} \left\lfloor \frac{n}{2} \right\rfloor \left\lfloor \frac{n-1}{2} \right\rfloor \left\lfloor \frac{n-2}{2} \right\rfloor \left\lfloor \frac{n-3}{2} \right\rfloor.$$

Equality is known to hold for $n \le 12$.

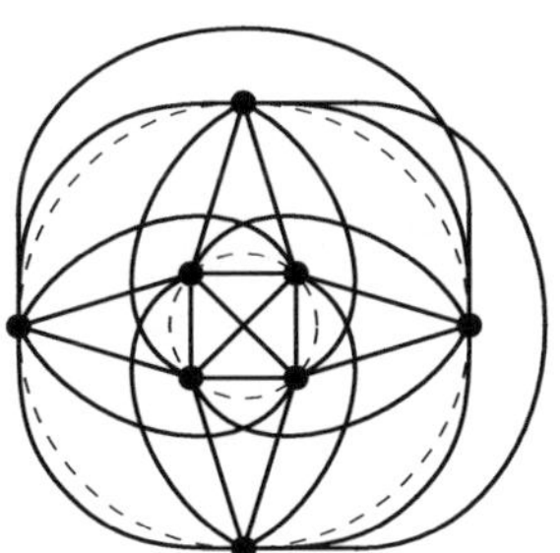

Figure 3.7. An optimal drawing of K_8

Hill also considered the *linear crossing number* $\overline{\operatorname{cr}}(G)$ of a graph G which is the fewest crossings in a drawing of G in which all of the edges are straight-line segments. In the light of Fáry's theorem, one might expect that this number is always equal to $\operatorname{cr}(G)$. In fact, as Daniel Bienstock and Nathaniel Dean have shown,[25] this is indeed the case for graphs with crossing number 3 or less, but there is no bound on the linear crossing number of graphs with crossing number 4.

Even for complete graphs the situation is somewhat surprising. Although $\operatorname{cr}(K_n)$ and $\overline{\operatorname{cr}}(K_n)$ are equal for $n \le 7$, $\overline{\operatorname{cr}}(K_8) = 19 = \operatorname{cr}(K_8) + 1$. However, equality holds again when $n = 9$ (both are 36), but there is inequality for all larger complete graphs. It is a bit surprising that, although the exact value of $\operatorname{cr}(K_n)$ is known only for $n \le 12$, $\overline{\operatorname{cr}}(K_n)$ is known up to $n = 27$ and also for $n = 30$ (where the value is 9726).

3.2.3 Products of cycles. The simplest bounded surface is a *sphere*; and for graphs, both planes and spheres serve as a basis. For most mathematicians, the next natural surface on which graphs might be drawn without crossings is probably a *torus,* or doughnut. Figure 3.8 shows a torus on which a grid is drawn.

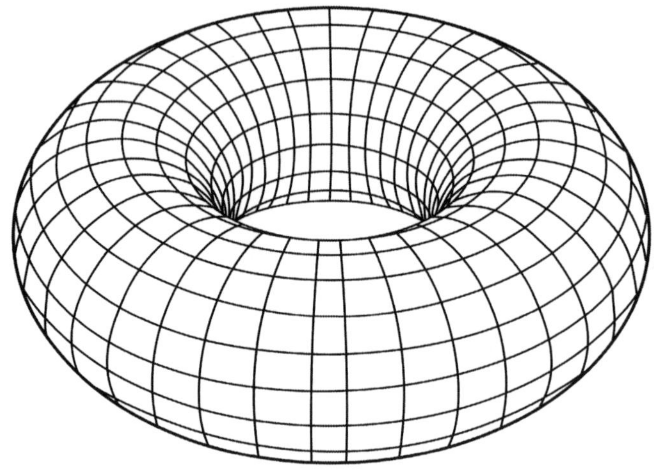

Figure 3.8. A doughnut with a grid

The graph of such a grid is the product of two cycles C_r and C_s (with $r, s \geq 3$). An elementary example is the graph $C_4 \times C_5$ shown in Figure 3.9, and it has crossing number 10 as illustrated.

A straightforward upper bound for the crossing number was conjectured to be equality.

Conjecture 3.1 (Product of cycles conjecture). *If $3 \leq r \leq s$, then the crossing number of $C_r \times C_s$ is $(r-2)s$.*

As is often the case, this conjecture was approached piece by piece, with the first result being for the two families $C_3 \times C_s$ and $C_4 \times C_s$. Further results were achieved with the passage of time, including the following theorem.

Theorem 3.5 (Product of cycles theorem). *If $3 \leq r \leq 7$, then $\mathrm{cr}(C_r \times C_s) = (r-2)s$ for all s.*

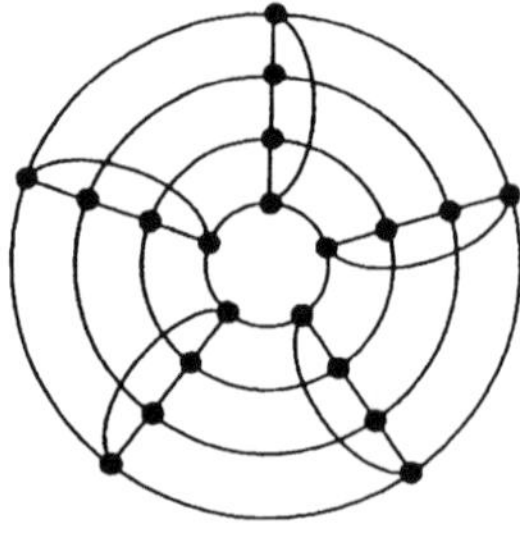

Figure 3.9. An optimal drawing of $C_4 \times C_5$

A different approach, with different results was made by Lev Glebsky and Gelasio Salazar in a series of papers spread over many years.

Theorem 3.6 (Glebsky and Salazar's theorem). *If $r \geq 3$ and $s \geq r(r+1)$, then* $\mathrm{cr}(C_r \times C_s) = (r-2)s$.

The impact of this theorem is interesting as there are values of r and s with $r \leq s < r(r+1)$ for which the crossing number $\mathrm{cr}(C_r \times C_s)$ has not yet been determined!

3.2.4 Thrackles. The pentagram in Figure 3.10 is (in a sense) the opposite of a plane embedding, being a drawing in which every two nonadjacent edges cross exactly once.

Figure 3.10. A 5-cycle with all edges either adjacent or crossing

One of the oldest problems of this type can be traced back to 1885, when Richard Baltzer[26] asked for the largest number of crossings in a rectlinear plane drawing of an n-gon. Here, the solution for odd polygons resembles that of the 5-cycle, with all nonadjacent edges crossing, but the even case does not. In 1923 Ernst Steinitz[27] proved that the maximum number of crossings for an n-cycle in the plane is $\frac{1}{2}n(n-3)$ if n is odd and $\frac{1}{2}n(n-4)+1$ if n is even. Figure 3.11 shows drawings of C_7 with 14 crossings and C_8 with 17 crossings.

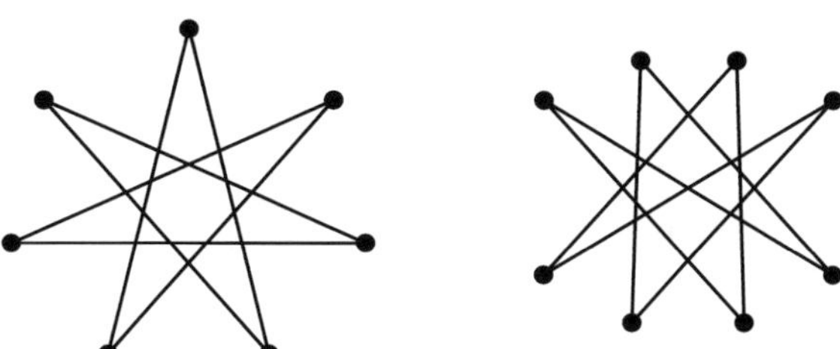

Figure 3.11. Maximizing the number of crossings in polygons

If we drop the requirement for straight-line edges, then the corresponding question was answered for cycles by Heiko Harborth,[28] who proved that for $n \geq 5$ the maximum crossing number of the cycle C_n is $\frac{1}{2}n(n-3)$. Harborth also solved the corresponding problem for complete and complete bipartite graphs: the maximum crossing number of K_n is $\binom{n}{4}$ and of the complete bipartite graph $K_{r,s}$ is $\binom{r}{2}\binom{s}{2}$. Unlike the situation for cycles, these values can be achieved by drawings in which the edges are straight-line segments.

A variant of this problem that also involves many crossings requires *all pairs* of nonadjacent edges to cross. Classic examples are the 5-cycle in Figure 3.10 and the 6-cycle in Figure 3.12. The renowned mathematician John Conway[29] called such drawings *thrackles,* and the following conjecture, which remains open, was announced at the first British Combinatorial Conference in Oxford in July 1969.[30]

Conjecture 3.2 (Conway's thrackle conjecture). *If G is a connected graph drawn in the plane with all pairs of nonadjacent edges crossing, then it has at least as many vertices as edges.*

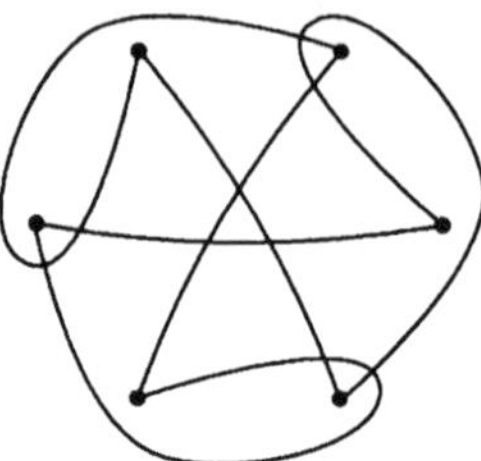

Figure 3.12. A 6-cycle with the maximum number of crossings

3.3 Graphs on higher surfaces

We begin our discussion of graphs on surfaces in general by clarifying what we mean by a surface. In technical terms, a *surface* is a compact 2-manifold—each point on the surface has a neighborhood that is homeomorphic to an open disk, and the surface is closed and bounded. Note that this definition is quite restrictive: a plane does not qualify as a surface because it is not bounded, and a Möbius band does not qualify because it has a boundary. Except in the simplest cases, surfaces can usefully be represented as polygons with pairs of sides identified in prescribed directions.

In general, a surface is *orientable* if we can consistently choose a positive sense of rotation (say, clockwise) around all points: the simplest orientable surface, other than a sphere, is a torus (as indicated by the "doughnut" in Figure 3.8). Figure 3.13 shows a torus and its representation as a rectangle with opposite sides identified.

Likewise, a surface is *nonorientable* if we cannot consistently choose a positive sense of rotation around all points; the simplest nonorientable surfaces are a projective plane and a Klein bottle. A Klein bottle and its rectangular representation are shown in Figure 3.14.

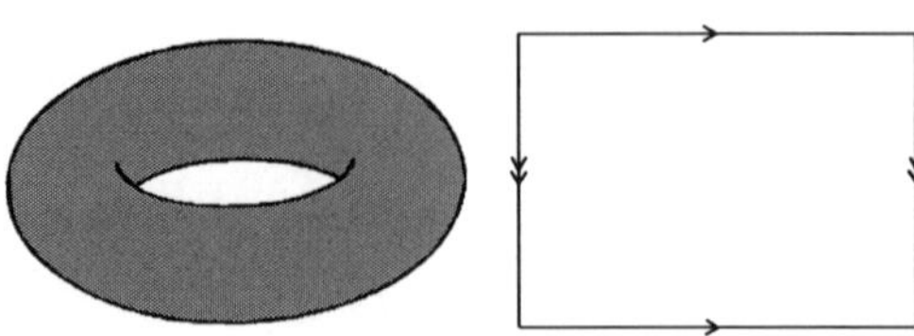

Figure 3.13. A torus

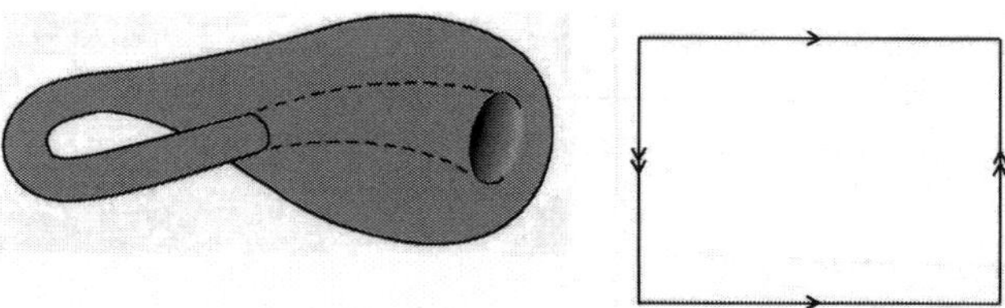

Figure 3.14. A Klein bottle

A surface S can be modified by adding a handle to it. To do so, we remove two disjoint open disks from S and then identify their boundaries with the ends of a cylinder in a consistent manner. Similarly, S can be modified by adding a cross-cap. To do so, we remove one open disk from S and identify its boundary with that of a Möbius band: this is equivalent to identifying opposite points on the boundary of the disk (see Figure 3.15).

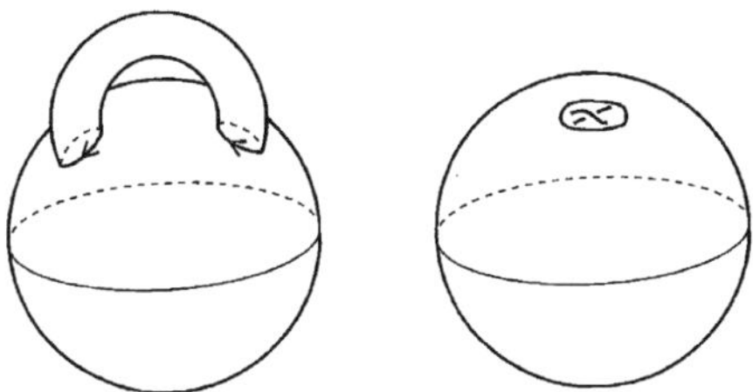

Figure 3.15. Adding a handle and adding a cross-cap

If we take a sphere and add $h(\geq 1)$ handles in any way whatever, the result is always topologically the same—we denote the resulting orientable surface by S_h (see Figure 3.16 for two drawings of the surface S_2). Similarly, if we take a sphere and add $k(\geq 1)$ cross-caps in any way whatever, the result is always topologically the same— we denote the resulting nonorientable surface by N_k; for example, a sphere with one cross-cap is a projective plane N_1 and a sphere with two cross-caps is a Klein bottle N_2.

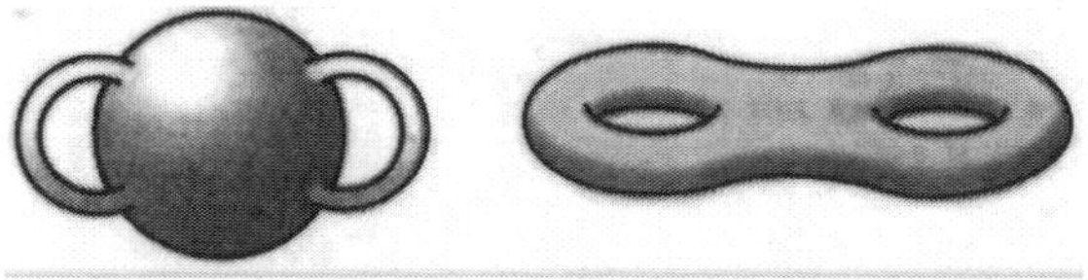

Figure 3.16. Two views of a double torus

In 1923, the American mathematician H. Roy Brahana proved a remarkable result that classifies all surfaces.[31]

Theorem 3.7 (Brahana's classification theorem). *Every surface is topologically the same as either an orientable surface S_h with $h \geq 0$, or a nonorientable surface N_k with $k \geq 1$. Furthermore, for all $h \geq 0$ and all $k \geq 1$, a sphere with h handles and k cross-caps is nonorientable and is topologically the same as N_{2h+k}.*

Many concepts for drawings and embeddings on a plane carry over in a natural way to these other surfaces. In particular, a graph is *embeddable* on a surface S if it can be drawn on S with no edges crossing, and an embedding is *cellular* if every region is homeomorphic to an open disk. Figure 3.17 shows cellular embeddings of the complete graphs K_6 on a projective plane and K_7 on a torus, where the projective plane is presented as a circular disk with diametrically opposite points identified, whereas the torus is presented as before as a rectangle with opposite sides identified. We observe that in both of these embeddings every region is triangular.

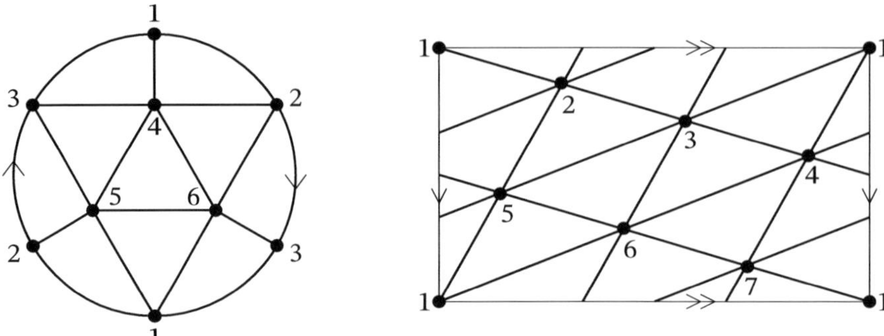

Figure 3.17. Cellular embeddings of K_6 on a projective plane and K_7 on a torus

One of the key results on graph embeddings is that every surface has its own version of Euler's formula; as mentioned in Chapter 1, the version for orientable surfaces was due to Lhuilier in 1812–13.

Formula 3.1 (Euler's formula for surfaces). *If a surface S has a cellular embedding of a connected graph G with n vertices, m edges, and r regions, then*

$$n - m + r = 2 - 2h \text{ if } S = S_h, \quad \text{and} \quad n - m + r = 2 - k \text{ if } S = N_k.$$

The number $2-2h$ or $2-k$ associated with a surface is called its *Euler characteristic*; for example, a torus S_1 ($h = 1$) and a Klein bottle N_2 ($k = 2$) both have Euler characteristic 0, and a projective plane N_1 ($k = 1$) has Euler characteristic 1. A sphere with h handles and k cross-caps has Euler characteristic $2 - 2h - k$. Brahana's result then says that two surfaces are topologically the same if and only if they are both orientable or both nonorientable and they have the same Euler characteristic.

3.3.1 The Heawood conjecture.
As we noted in Chapter 2, in his 1890 paper pointing out the flaw in Kempe's purported proof of the four-color theorem, Percy Heawood initiated the study of map colorings on surfaces other than the plane.[32] In particular, he demonstrated that some maps on a torus require seven colors, and he proved that no map needs more than this.

Heawood actually did more, although less than he claimed. He found a remarkable bound for the maximum number of colors that are needed to color the regions of any map drawn on a given orientable surface, with neighboring regions colored differently. In particular, he showed that the regions of every embeddable map on the surface S_h

(for $h \geq 1$) can be colored with at most

$$\left\lfloor \tfrac{1}{2}(7 + \sqrt{48h + 1}) \right\rfloor$$

colors. A simple calculation shows the important fact, used below, that for a given number n,

$$7 \leq n \leq \tfrac{1}{2}(7 + \sqrt{48h + 1}) \text{ is equivalent to } 1 \leq \tfrac{1}{12}(n - 3)(n - 4) \leq h.$$

However, although the number of colors given by Heawood's formula is sufficient for each such surface S_h, it is far from obvious in general that there must be maps that actually require this number of colors, as happens for a torus. Heawood supposed this to be the case, but his proof was deficient and his belief became known as the *Heawood conjecture*: it remained unproved until the 1960s, as we shall see. Heawood admitted the deficiency and went on to write several further papers on the problem.

Conjecture 3.3 (Heawood conjecture). *For $h \geq 1$, every map that is drawn on the surface S_h can be colored with*

$$\left\lfloor \tfrac{1}{2}(7 + \sqrt{48h + 1}) \right\rfloor$$

colors, and there are maps on S_h that require this number of colors.

In 1891, the German mathematician Lothar Heffter pointed out the gap in Heawood's argument.[33] He then asked for the smallest possible number of handles on an orientable surface on which we can draw n mutually neighboring regions; for example, on a sphere (where $h = 0$) we can draw four neighboring regions but not five, and on a torus we can draw seven neighboring regions but not eight (see Figure 3.18).

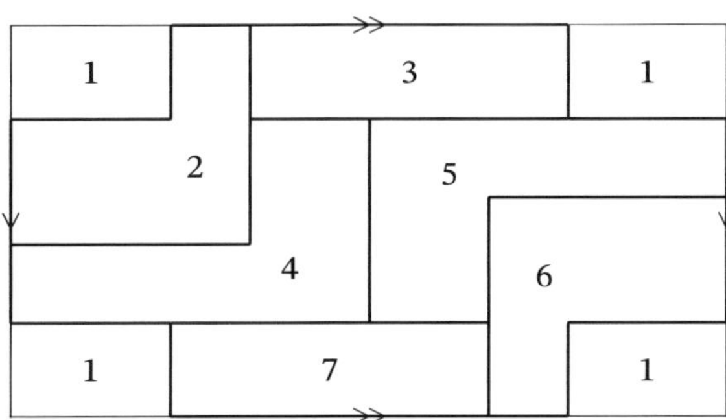

Figure 3.18. Seven neighboring regions on a torus

Heffter next dualized the problem to that of finding the smallest number h of handles on an orientable surface S_h on which we can embed the complete graph K_n, and observed that

$$h \geq \left\lceil \tfrac{1}{12}(n - 3)(n - 4) \right\rceil,$$

which follows directly from Heawood's formula, as noted above. He then proved that this lower bound for h gives the correct value for $n \leq 12$ and for some other values; for example, when $n = 7$ the formula yields $h = 1$, a torus.

To prove that Heawood's formula always gives the correct number of colors, it is sufficient to prove equality in Heffter's formula; thus, for the torus, it is sufficient to note that K_7 can be embedded there. But this is also necessary: any graph on the torus requiring seven colors must contain K_7. This was first realized by Peter Ungar in the 1940s, and Gabriel Dirac[34] subsequently proved that this is so in general (except for the sphere): Heffter's lower bound for the genus of a complete graph is achieved if and only if Heawood's upper bound for the number of colors needed for embedded graphs is achieved.

Given a surface S, the maximum chromatic number of a graph that is embeddable on S is called the *chromatic number* $\chi(S)$. Proving equality in Heffter's inequality is then not only sufficient, but is also necessary, to prove the *map color theorem.*

Theorem 3.8 (Orientable map color theorem). *For $h \geq 1$, the chromatic number of the orientable surface S_h is*

$$\chi(S_h) = \left\lfloor \tfrac{1}{2}(7 + \sqrt{48h + 1}) \right\rfloor.$$

Until 1910, all attempts to extend map coloring to surfaces had concentrated on orientable surfaces. But in that year, the Austrian mathematician Heinrich Tietze wrote a note on coloring maps on a Möbius band and a projective plane, showing that every map drawn on either of them can be colored with six colors.[35] He also provided analogues for nonorientable surfaces of the formulas of Heawood and Heffter.

3.3.2 The genus of a graph. The *orientable genus* $\gamma(G)$ of a graph G is the minimum number of handles that must be added to a sphere for G to be embeddable thereon. As we have seen, equality in Heffter's inequality for $\gamma(K_n)$ implies the orientable map color theorem. The solution was finally completed in 1968 by the German mathematician Gerhard Ringel and the American J. W. T. Youngs,[36] while Ringel was on sabbatical leave in California with Youngs. It was described in detail in the 1974 book on coloring graphs on surfaces by Ringel, mentioned in the Further Reading section.

Theorem 3.9 (Ringel–Youngs theorem). *For $n \geq 3$, the orientable genus of the complete graph K_n is*

$$\gamma(K_n) = \left\lceil \tfrac{1}{12}(n - 3)(n - 4) \right\rceil.$$

Because of the form of the answer, it had transpired that the problem splits naturally into twelve cases, corresponding to the various values of n (modulo 12). Particularly important were those complete graphs for which the expression inside the brackets is an integer—that is, when $n \equiv 0, 3, 4,$ or $7 \pmod{12}$.

In order to facilitate the construction of the embeddings, Ringel found it helpful to create "current graphs": these are directed graphs with labels on the arcs that satisfy Kirchhoff's current law from the theory of electrical networks. These ingenious devices provided just what was needed. For a further discussion of Ringel and Youngs's proof of the Heawood conjecture, see the survey by Arthur White, which summarizes the twelve cases into which the proof is divided, or the treatment in *Graph Theory in America.*[37]

Subsequently, substantial simplifications arose when Jonathan Gross and Thomas Tucker developed dual mechanisms called "voltage graphs".[38] To achieve complicated graph embeddings, one can use simpler ones with "voltages" assigned to the edges that satisfy Kirchhoff's voltage law.

An embedding of a graph on an orientable surface can be described purely combinatorially by a *rotation system*, consisting of a clockwise ordering of the edges around each vertex. For example, in the table in Figure 3.19, the list of eight vertices with their incident edges is a rotation system that results in the given embedding of $K_{4,4}$ on a torus.

vertex	edges clockwise
s	sx, sw, sz, sy
t	tx, ty, tz, tw
u	ux, uw, uz, uy
v	vx, vy, vz, vw
w	ws, wt, wu, wv
x	xs, xv, xu, xt
y	ys, yt, yu, yv
z	zs, zv, zu, zt

Figure 3.19. Embedding $K_{4,4}$ on a torus, based on a rotation system of its edges

There is a one-to-one correspondence between cellular embeddings and such rotation systems, consisting (for each vertex) of a circular permutation of the edges that are incident to that vertex. This had been implicit in Heffter's paper, but was treated explicitly in 1960 by Jack Edmonds[39] and described in his master's thesis at the University of Maryland; it was also already used in the 1959 book on coloring graphs on surfaces by Ringel, mentioned in the Further Reading section, before being developed further by Youngs and Ringel. Edmonds's treatment provided a method for finding the number of handles of the surface of the embedding from the rotation system. Finding the minimum number of handles h for which the graph can be embedded on S_h is thus equivalent to finding certain such rotation systems, and current and voltage graphs are useful devices for describing these systems; but no efficient algorithm is known for determining the orientable genus of a given graph, not even for cubic graphs.

The nonorientable analogue of the genus of a graph G is the *nonorientable genus* (or *cross-cap number*) $\gamma'(G)$, the minimum number of cross-caps that must be added to a sphere for G to be embeddable thereon. It follows from the embeddability of K_6 on a projective plane (see Figure 3.19) that $\gamma'(K_6) = 1$. However, although a Klein bottle has the same Euler characteristic as a torus, and K_7 can be embedded on a torus, it cannot be embedded on a Klein bottle, a fact that was first proved by Philip Franklin in 1934.[40]

Sixteen years before the genus of all complete graphs was found, Ringel had already established all of their cross-cap numbers.[41] It turns out that K_7 is indeed exceptional, being the only complete graph whose cross-cap number does not equal the bound arising from Euler's formula.

Theorem 3.10 (Ringel's theorem). *For $n \geq 3$, the nonorientable genus of the complete graph K_n is*

$$\gamma'(K_n) = \left\lceil \tfrac{1}{6}(n-3)(n-4) \right\rceil, \text{ except that } \gamma'(K_7) = 3.$$

Extending the fact that K_7 cannot be embedded on a Klein bottle, Franklin showed that every graph that can be embedded thereon is 6-colorable. Furthermore, just as the

Ringel–Youngs theorem yields what is needed to determine the chromatic number of all orientable surfaces other than a sphere, so this earlier theorem of Ringel yields the chromatic number of all nonorientable surfaces, again with the notable exceptional case of a Klein bottle.

Theorem 3.11 (Nonorientable map color theorem). *For $k \geq 1$, the chromatic number of the nonorientable surface N_k is*

$$\chi(N_k) = \left\lfloor \tfrac{1}{2}(7 + \sqrt{24k + 1}) \right\rfloor,$$

except that $\chi(N_2) = 6$.

3.3.3 Kuratowski-type theorems.

Earlier we presented Kuratowski's theorem, that a graph is can be embedded on a sphere if and only if it has no subgraph homeomorphic to either of the two "forbidden subgraphs" K_5 and $K_{3,3}$, and it is natural to ask whether there are corresponding results for other surfaces. It turns out that every surface has a finite family of such forbidden subgraphs (as we shall explain below), but little is known about them except in a couple of specific cases. In fact, other than a sphere, the only surface for which such a family is known explicitly is a projective plane; here, a graph can be embedded if and only if it has no subgraph homeomorphic to one of 103 specified graphs, as shown by Henry Glover and Philip Huneke in 1978.[42]

We also saw a variation on Kuratowski's theorem by Wagner, who proved that a graph is planar if and only if it has no subgraph that is contractible to K_5 or $K_{3,3}$, and we might ask whether there are corresponding results for other surfaces. To answer this, it is convenient to define a *minor* of a graph G to be any graph that can be obtained from G by deleting vertices and deleting and contracting edges. Wagner's theorem then says that a graph is planar if and only if it does not contain K_5 or $K_{3,3}$ as a minor.

To investigate such questions, we define a *minor-minimal forbidden family $M(S)$* for a surface S to be a set of graphs with the following properties:

- no graph in $M(S)$ can be embedded in S;

- every graph that cannot be embedded in S has a graph in $M(S)$ as a minor;

- no graph in $M(S)$ is a minor of another graph in $M(S)$.

It follows that K_5 and $K_{3,3}$ form a minor-minimal forbidden family for a sphere. For a projective plane, Glover and Huneke found a minor-minimal family containing just 35 graphs.

One of the foremost results in graph theory is the following spectacular theorem on graph minors, announced by Neil Robertson and Paul Seymour in 1984.[43]

Theorem 3.12 (Robertson–Seymour theorem). *Every infinite collection of graphs contains at least one graph that is a minor of another member of the family.*

A classic topological consequence is the following.

Corollary 3.1. *The family of minor-minimal forbidden graphs of every surface is finite.*

It follows from this corollary that, for every surface, the set of homeomorphically minimal forbidden graphs is also finite, a result that had been conjectured by Dénes

König in 1936,[44] and was independently established in 1989 for general nonorientable surfaces by Archdeacon and Huneke.[45] A simpler proof, although still related to results of Robertson and Seymour, was published by Carsten Thomassen in 1997.[46]

The publication of a long and involved proof of Robertson and Seymour's theorem was completed in 2004,[47] thereby settling a long-standing conjecture of Wagner. Indeed, Wagner had considered infinite families of finite graphs, and in his book of 1970 he had already given an example of such a family where no two graphs are homeomorphic.[48] He had further asked for an infinite set of finite graphs where no graph is a minor of another. The Robertson–Seymour theorem shows that no such set can exist.

3.4 Conclusion

In this chapter we have concentrated mainly on classical results. Although topological graph theory may seem to be a somewhat restricted branch of graph theory, this is not so: the Robertson–Seymour theory on minors (to which we return in Chapter 8) uses topological ideas as significant tools for obtaining general results, so that topological graph theory has come to play a natural and important role in the general study of discrete mathematics. Conversely, combinatorial graph theory is useful for studying topics in topology, as shown in work of Carsten Thomassen and W. T. Tutte.[49]

Further Reading

The first books to discuss the drawing of graphs on surfaces seem to have been these monographs:

Dénes König, *The Elements of Analysis Situs* (in Hungarian), Budapest (1918);

André Sainte-Laguë, *Géométrie de Situation et Jeux, Mémorial des Sciences Mathématiques* 41, Gauthier-Villars (Paris) (1929).

The first book devoted specifically to coloring problems on graphs and surfaces was

G. Ringel, *Färbungsprobleme auf Flächen und Graphen*, VEB Deutscher Verlag der Wissenschaften (1959).

For more information on topological graph theory, see the following books:

Arthur T. White, *Graphs, Groups and Surfaces*, North-Holland (1973); revised edn. (1984);

Jonathan Gross and Thomas W. Tucker, *Topological Graph Theory*, Wiley (1987);

C. Paul Bonnington and Charles H. C. Little, *The Foundations of Topological Graph Theory*, Springer (1995);

Bojan Mohar and Carsten Thomassen, *Graphs on Surfaces*, Johns Hopkins Univ. Press (2001);

L. W. Beineke and R. J. Wilson (eds.), *Topics in Topological Graph Theory*, Cambridge Univ. Press (2009).

A paper with useful historical information about Kuratowski's theorem is
> John W. Kennedy, Louis V. Quintas, and Maciej M. Sysło, The theorem
> on planar graphs, *Historia Mathematica* 12 (1985), 356–68.

A detailed discussion of Ringel and Youngs's proof of the Heawood conjecture appears in
> G. Ringel, *Map Color Theorem*, Springer (1974).

Whitney's writings on graph theory appear in
> Hassler Whitney, *Collected Works*, Vol. 1, Birkhäuser (1992).

A substantial book on crossing numbers is
> M. Schaefer, *Crossing Numbers of Graphs*, CRC Press (2018).

Notes and References

[1]C. Kuratowski, Sur le problème des courbes gauches en topologie, *Fund. Math.* 15 (1930), 271–283. His proof is based on point-set topological considerations, and an English translation was given by Jan Jawarowski, *Graph Theory, Lagow 1981*, Lecture Notes in Mathematics 1018, Springer (1983), 1–13.

[2]We note that, while proving his theorem, Kuratowski initially believed K_5 to be the only obstruction to planarity, and that $K_{3,3}$ turned up unexpectedly only as he was nearing the end of his argument: see K. Kuratowski, My personal recollections connected with the research on some topological problems, *Colloquio Internazionale sulle Theorie Combinatorie (Rome 1975)*, Vol. I, Atti dei Convegni Lincei 17, Accademia Nazionale dei Lincei (1976), 43–47.

[3]K. Menger, Über plättbare Dreiergraphen und Potenzen nichtplättbarer Graphen, *Anzeiger der Akademie der Wissenschaften in Wien* 67 (1930), 85–86.

[4]O. Frink and P. A. Smith, Abstract 179, *Bull. Amer. Math. Soc.* 36 (1930), 214.

[5]Whitney's paper on planar graphs, based on part of his doctoral thesis, was H. Whitney, Nonseparable and planar graphs, *Trans. Amer. Math. Soc.* 34 (1932), 339–362 = *H. Whitney Collected Papers*, Vol. I, Birkhäuser (1992), 37–60.

[6]These papers of Whitney are the following:
- The coloring of graphs, *Ann. Math.* 33 (1932), 688–718;
- Congruent graphs and the connectivity of graphs, *Amer. J. Math.* 54 (1932), 150–168;
- On the classification of graphs, *Amer. J. Math.* 55 (1933), 236–244;
- On the abstract properties of linear dependence, *Amer. J. Math.* 57 (1935), 509–533.

These appear in his *Collected Works*, Vol. 1, pp. 80–110, 61–79, 116–124, and 147–171.

[7]H. Whitney and W. T. Tutte, Kempe chains and the four color problem, *Util. Math.* 2 (1972), 241–281 = *Collected Works* Vol. 1, 185–225; reprinted in *Studies in Graph Theory* II (ed. D. R. Fulkerson), MAA (1975).

[8]K. Wagner, Über eine Eigenschaft der ebenen Komplexe, *Math. Ann.* 114 (1937), 570–590.

[9]F. Harary and W. T. Tutte, A dual form of Kuratowski's theorem, *Canad. Math. Bull.* 8 (1965), 17–20, 373.

[10]The two original articles on graphs whose edges are nonintersecting line segments are K. Wagner, Bemerkungen zum Vierfarbenproblem, *Jber. Deutsch. Math.-Verein.* 46 (1936), 26–32 and I. Fáry, On straight line representation of planar graphs, *Acta Univ. Szeged Sect. Sci. Math.* 11 (1948).

[11]W. Schnyder, Embedding planar graphs on the grid, *Proc. 1st ACM-SIAM Sympos. Discrete Algorithm* (1990), 138–148.

[12]The *union* $G_1 \cup G_2$ of two graphs G_1 and G_2 consists of those vertices and edges in at least one of the two graphs; the *intersection* $G_1 \cap G_2$ consists of those vertices and edges in both G_1 and G_2.

[13] For a history of biplanarity, see
L. W. Beineke, Biplanar graphs: A survey, *Comput. Math. Appl.* 34 (1997), 1–8.

[14] A. Mansfield, Determining the thickness of graphs is NP-hard, *Math. Proc. Cambridge Philos. Soc.* 93 (1983), 9–23.

[15] This was first reported by Martin Gardner in his mathematics column in *Scientific American*: M. Gardner, Mathematical games, *Sci. Amer.* 242 (2) (Feb. 1980), 14–19.

[16] D. Boutin, E. Gethner, and T. Sulanke, Thickness-two graphs, I. New nine-critical graphs, permuted layer graphs, and Catlin's graphs, *J. Graph Theory* 57 (2008), 198–214 and E. Gethner and T. Sulanke, Thickness-two graphs. II. More new nine-critical graphs, independence ratio, cloned planar graphs, and singly and doubly outerplanar graphs, *Graphs Combin.* 25 (2009), 197–217.

[17] J. Battle, F. Harary, and Y. Kodama, Every planar graph with nine points has a nonplanar complement, *Bull. Amer. Math. Soc.* 68 (1962), 569–571 and W. T. Tutte, The nonbiplanar character of the complete 9-graph, *Canad. Math. Bull.* 6 (1963), 319–330.

[18] W. T. Tutte, The thickness of a graph, *Nederl. Akad. Wetensch. Proc. = Indag. Math.* 25 (1963), 567–577.

[19] L. W. Beineke and F. Harary, On the thickness of the complete graph. *Bull. Amer. Math. Soc.* 70 (1964), 618–620. This was for graphs with more than 10 vertices, except for those with n vertices where $n \equiv 4 \pmod 6$. Jean Mayer showed that, unlike K_9 and K_{10}, K_{16} was not an exception to the general formula: (J. Mayer, Décomposition de K_{16} en trois graphes planaire, *J. Combin. Theory (B)* 13 (1972), 71). A few years later, the proof was completed using the earlier techniques: V. B. Alekseev and V. S. Gončakov, The thickness of an arbitrary complete graph (Russian), *Mat. Sb.* (N.S.) 101(143) (1976), 212–230; (English translation: *Math. USSR-Sb.* 30 (1976), 187–202).

[20] L. W. Beineke, F. Harary, and J. W. Moon, On the thickness of the complete bipartite graph, *Proc. Cambridge Philos. Soc.* 60 (1964), 1–5.

[21] P. Turán, A note of welcome, *J. Graph Theory* 1 (1977), 7–9. This note was the introduction to the first issue of this journal.

[22] K. Zarankiewicz, On a problem of P. Turán concerning graphs, *Fund. Math.* 41 (1954), 137–145.

[23] R. K. Guy, A combinatorial problem, *Bull. Malayan Math. Soc. (Nabla)* 7 (1960), 68–72.

[24] Information about Anthony Hill and his notes on the crossing numbers of complete graphs can be found in L. Beineke and R. Wilson, The early history of the brick factory problem, *Math. Intelligencer* 32 (2010), 41–48. Guy and Hill later collaborated on a related crossing number problem: R. K. Guy and A. Hill, The crossing number of the complement of a circuit, *Discrete Math.* 5 (1973), 335–344.

[25] D. Bienstock and N. Dean, Bounds for rectilinear crossing numbers, *J. Graph Theory* 17 (1993), 333–348.

[26] R. Baltzer, Eine Errinerung an Möbius und seinen Freund Weiske, *Berichte über die Verhandlungen der Königlich Sächsischen Gesellschaft der Wissenschaften zu Leipzig* 37 (1885), 1–6.

[27] E. Steinitz, Über die Maximalzahl der Doppelpunkte bei ebenen Polygonen von gerader Seitenzahl, *Math. Z.* 17 (1923), 116–129.

[28] H. Harborth has written several papers on graphs in the plane, two of which are of particular interest here:
  • H. Harborth, Parity of numbers of crossings for complete n-graphs, *Math. Slovaka* 26 (1976), 77–95;
  • H. Harborth, Drawings of the cycle graph, *Congr. Numer.* 66 (1988), 15–22.
For more on this topic, see the book by Schaefer mentioned above.

[29] According to Dan Archdeacon, John Conway introduced the term "thrackle" with this explanation:
"When I was a teenager, on holiday with my parents in Scotland, we once stopped to ask directions of a man who was fishing by the side of a lake. He happened to mention that his line was thrackled…"

[30] D. R. Woodall, Thrackles and deadlock, *Combinatorial Mathematics and its Applications* (ed. D. J. A. Welsh), Academic Press (1971), 335–347. The best known bound on Conway's conjecture is in L. Lovász, J. Pach, and M. Szegedy, On Conway's thrackle conjecture, *Discrete Comput. Geom.* 18 (1997), 369–376.

[31] H. R. Brahana, Systems of circuits of two-dimensional manifolds, *Ann. of Math.* 30 (1923), 234–243.

[32] P. J. Heawood, Map-colour theorem, *Quart. J. Pure Appl. Math.* 24 (1890), 332–338.

[33] L. Heffter, Über das Problem der Nachbargebiete, *Math. Ann.* 38 (1891), 477–508.

[34] G. A. Dirac, Map-colour theorem, *Canad. J. Math.* 4 (1952), 480–490.

[35] H. Tietze (1910), Einige Bemerkungen über das problem des Kartenfärbens auf einseitigen Fläschen, *Jber. Deutsch. Math.-Verein.* 19 (1910), 155–159.

[36] G. Ringel and J. W. T. Youngs, Solution of the Heawood map-coloring problem, *Proc. Nat. Acad. Sci. U.S.A.* 60 (1968), 438–445; fuller details appear in Ringel's 1974 book *Map Color Theorem* (see above).

[37] A. T. White, The proof of the Heawood conjecture, *Selected Topics in Graph Theory* (ed. L. W. Beineke and R. J. Wilson), Academic Press (1978), 51–81: this chapter summarizes the twelve cases into which Ringel and Youngs's proof is divided. R. Wilson, J. J. Watkins, and D. J. Parks, *Graph Theory in America*, Princeton Univ. Press (2023).

[38] J. L. Gross and T. W. Tucker, Quotients of complete graphs, revisiting the Heawood map-coloring theorem, *Pacific J. Math.* 55 (1974), 391–402.

[39] J. R. Edmonds, A combinatorial representation for polyhedral surfaces, *Notices Amer. Math. Soc.* 7 (1960), 646.

[40] P. Franklin, A six color problem, *J. Math. Phys.* 14 (1934), 228–231.

[41] G. Ringel, Farbensatz für nichtorientierbare Flächen beliebigen Geschlechts, *J. Reine Angew. Math.* 190 (1952), 129–147.

[42] H. Glover and J. P. Huneke, The set of irreducible graphs for the projective plane is finite, *Discrete Math.* 22 (1978), 243–256.

[43] N. Robertson and P. D. Seymour, Generalizing Kuratowski's theorem, *Congr. Numer.* 45 (1984), 129–138.

[44] D. König, *Theorie der endlichen und unendlichen Graphen*, Akademische Verlagsgesellschaft (1936); this book is generally regarded as "the first book on graph theory".

[45] D. Archdeacon and P. Huneke, A Kuratowski theorem for nonorientable surfaces, *J. Combin. Theory (B)* 46 (1989), 173–231.

[46] C. Thomassen, A simpler proof for the excluded minor theorem for higher surfaces, *J. Combin. Theory (B)* 70 (1997), 306–311.

[47] N. Robertson and P. Seymour, Graph minors XX. Wagner's conjecture, *J. Combin. Theory (B)* 35 (2004), 39–61.

[48] K. Wagner, *Graphentheorie*, BI Hochschultaschenbücher (1970).

[49] See the book by Mohar and Thomassen mentioned above, and also these papers:

• C. Thomassen, The Jordan-Schönflies theorem and the classification of surfaces, *Amer. Math. Monthly* 99 (1992), 116–130;

• W. T. Tutte, *The Combinatorial Sphere*, Lecture Notes, University of Waterloo (1973).

4

Graphs, Linear Algebra, and Groups

The origins of Chapter 2 (*Coloring Maps and Graphs*) can be traced back to the four-color problem which dates from 1852, whereas those of Chapter 3 (*Graphs on Surfaces*) lie in the papers of Percy Heawood and Lothar Heffter from the 1890s. We now turn our attention to a subject that also had its origins in the 19th century, with the development of linear algebra and the theory of groups, but where the main applications to graph theory did not begin until the 20th century.

Algebraic graph theory can be broadly divided into two distinct areas. In the first half of this chapter we explore the use of ideas from linear algebra, and the algebra of matrices in particular, to investigate the properties of graphs; especially important here are the adjacency and incidence matrices of a graph, and ways in which the eigenvalues of the adjacency matrix give us information about the corresponding graph, and vice versa. In the second part of this chapter we survey the connections between the symmetries of graphs and the algebraic properties of their automorphism groups.

The two parts of this chapter are not unrelated—for example, the automorphism group of a graph can be regarded as the set of all permutation matrices that commute with the graph's adjacency matrix. However, as we shall see, the milestones in the first half of this chapter have a distinctly different flavor from those in the second half. This chapter assumes some knowledge of linear algebra and group theory.

4.1 Linear algebra

If G is a graph, not necessarily simple, with its vertices labeled $1, 2, \ldots, n$, then its *adjacency matrix* $\mathbf{A}$ is the $n \times n$ matrix whose (i, j)-entry is the number of edges joining vertex i and vertex j. If, in addition, the edges are labeled $1, 2, \ldots, m$, then its *incidence matrix* $\mathbf{M}$ is the $n \times m$ matrix whose (i, j)-entry is 1 if vertex i is incident with edge j, and is 0 otherwise. Figure 4.1 shows a labeled graph G with its adjacency and incidence matrices.

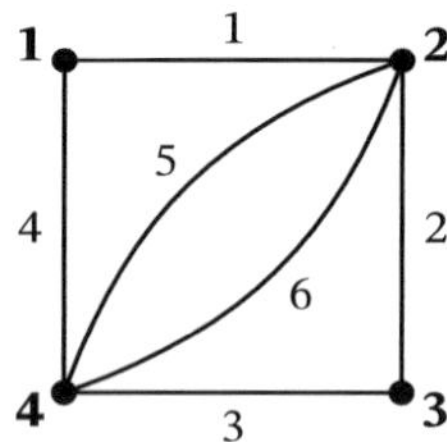

$$A = \begin{pmatrix} 0 & 1 & 0 & 1 \\ 1 & 0 & 1 & 2 \\ 0 & 1 & 0 & 1 \\ 1 & 2 & 1 & 0 \end{pmatrix} \qquad M = \begin{pmatrix} 1 & 0 & 0 & 1 & 0 & 0 \\ 1 & 1 & 0 & 0 & 1 & 1 \\ 0 & 1 & 1 & 0 & 0 & 0 \\ 0 & 0 & 1 & 1 & 1 & 1 \end{pmatrix}$$

Figure 4.1. A labeled graph and its adjacency and incidence matrices

4.1.1 Incidence matrices.

4.1.1 Incidence matrices. Early in the 20th century, a new *Encyklopädie der Mathematische Wissenschaften* (Encyclopedia of the Mathematical Sciences), designed to survey current knowledge in the subject, was published in Germany. Among its articles was one by Max Dehn of Germany and Poul Heegaard of Denmark on *Analysis Situs*, the study of position.[1] Following on from ideas of Kirchhoff on electrical networks and of Johann Benedict Listing and Henri Poincaré on complexes (see Chapter 1), their article opened with an algebraic treatment of "Liniensysteme" (line systems, or graphs), regarded as one-dimensional complexes constructed from 0-cells (vertices) and 1-cells (edges).

Oswald Veblen and the four-color problem. In 1912, building on Dehn and Heegaard's ideas, Oswald Veblen of Princeton University expressed the ideas of graph theory, and of planar graphs and maps in particular, in algebraic form.[2] Given a map with vertices labeled from 1 to α_0, edges labeled from 1 to α_1, and regions (countries) labeled from 1 to α_2, he defined two incidence matrices **A** and **B** with entries 0 and 1. In matrix **A**, the rows correspond to the vertices and the columns correspond to the edges. In matrix **B**, the rows correspond to the edges and the columns correspond to the regions, with the entry in row i and column j being 1 if edge i borders region j, and 0 otherwise. As an illustration, Veblen presented the map in Figure 4.2 and its associated matrices.

With each matrix, he then associated two systems of linear equations. For the vertex-edge incidence matrix **A**, one system assigns a variable x_i equal to 0 or 1 to each edge (column) i, and the equations have the form $x_i + x_j + \ldots = 0$ for each vertex (row). For example, for the matrix **A** in Figure 4.2 these equations are

$$x_1 + x_2 + x_3 = 0, x_2 + x_4 + x_5 = 0, x_1 + x_4 + x_6 = 0, \text{ and } x_3 + x_5 + x_6 = 0,$$

and the calculations are carried out in the finite field GF(2) where $1 + 1 = 0$. Here, one solution is $x_1 = x_3 = x_4 = x_5 = 1$ and $x_2 = x_6 = 0$, corresponding to the cycle with edges 1, 3, 4, and 5. Veblen showed more generally that in any solution to the equations, the edges labeled 1 always form an edge-disjoint set of cycles in the graph, from which he deduced that the number of linearly independent solutions is $\alpha_2 - 1$.

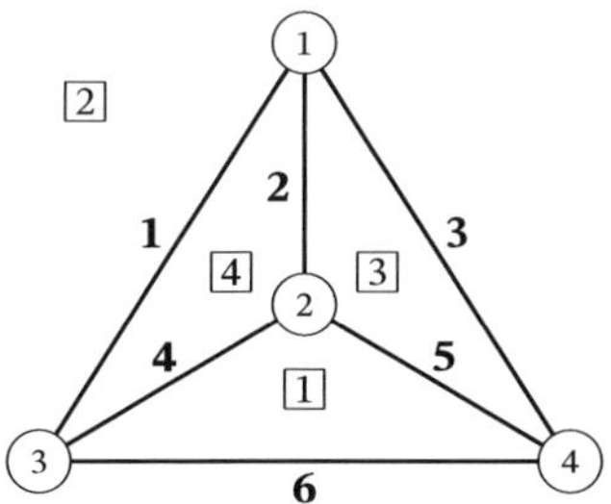

$$
\mathbf{A} = \begin{pmatrix} 1 & 1 & 1 & 0 & 0 & 0 \\ 0 & 1 & 0 & 1 & 1 & 0 \\ 1 & 0 & 0 & 1 & 0 & 1 \\ 0 & 0 & 1 & 0 & 1 & 1 \end{pmatrix} \qquad \mathbf{B} = \begin{pmatrix} 0 & 1 & 0 & 1 \\ 0 & 0 & 1 & 1 \\ 0 & 1 & 1 & 0 \\ 1 & 0 & 0 & 1 \\ 1 & 0 & 1 & 0 \\ 1 & 1 & 0 & 0 \end{pmatrix}
$$

Figure 4.2. A labeled map and its two incidence matrices

His second system of equations for the matrix $\mathbf{A}$ assigns a variable v_i to each vertex, and the equations have the form $v_i + v_j = 0$ for each edge $v_i v_j$. He then showed that the rank of $\mathbf{A}$ is $\alpha_0 - 1$, and it follows that $\alpha_2 - 1 = \alpha_1 - (\alpha_0 - 1)$, and hence that $\alpha_0 - \alpha_1 + \alpha_2 = 2$, which is Euler's formula.

Veblen then repeated the above process for the edge-region matrix $\mathbf{B}$. One system, arising from the rows of $\mathbf{B}$, has a variable y_i for each region and an equation for each edge of the form $y_i + y_j = 0$ for the two regions meeting along the edge. The other system, arising from the columns of $\mathbf{B}$, has a variable e_i for each edge and an equation of the form $e_i + e_j + \ldots = 0$ for the edges surrounding a region.

Having established his four systems of linear equations, all four satisfying $x = -x$, Veblen turned his attention to the four-color problem. Using the four elements $0, 1, i$, and $i+1$ to represent the colors, where $i^2 + i + 1 = 0$ and the calculations are carried out in the finite field GF(4) of four elements, he observed that a solution to the four-color problem consists in finding a set of values $(y_1, y_2, \ldots, y_{\alpha_2})$ that does not satisfy any of the equations $y_i + y_j = 0$ above.

Veblen's observation was that there is a 4-coloring of the regions of a planar map if and only if there is a vector y with coordinates from GF(4) for which $\mathbf{B}y$ has no coordinate equal to 0. The vector y may be written as $y_1 + iy_2$ where the coordinates of y_1 and y_2 are 0 or 1, and if we assume that the regions are not 2-colorable, then $\mathbf{B}y_2$ has a zero coordinate but the corresponding coordinate in $\mathbf{B}y_1$ is then not zero. Veblen showed that there are several such nonzero coordinates.

Now let C be a geometrical ring of r regions, where r is odd, in which each successive region has a boundary edge in common with its predecessor, and consider the subset of r equations $\mathbf{B}_C y = 0$ from $\mathbf{B} \cdot y$, corresponding to the r common boundary edges in the ring. Let the solutions over GF(4) to the restricted set be denoted by S_C. Veblen observed that a 4-coloring implies the existence of a y_1 such that $\mathbf{B}_C y_1$ has a nonzero coordinate, and he argued that the existence of such a variable y_1 for which

$\mathbf{B}_C y_1$ is nonzero for every possible C is both necessary and sufficient for a 4-coloring of the regions.

Theorem 4.1 (Veblen's theorem). *A map can be colored with four colors if and only if there exists a vector with coordinates 0 or 1 that does not lie in any subspace S_C, where C is a ring with an odd number of regions.*

Veblen's paper concluded with further algebraic formulations of the four-color problem. One was related to coloring the boundary edges of a cubic map with three colors, as considered by P. G. Tait in 1880 (see Chapter 2). Another involved a set of congruences that Percy Heawood had introduced in 1898: the four-color theorem is equivalent to the statement that there is a $\{1, -1\}$-labeling of the vertices for which the sum of the labels around each region is divisible by 3.

Following the appearance of this paper, Veblen was invited to give the 1916 American Mathematical Society Colloquium series of lectures on linear graphs, and these were subsequently published in 1922.[3]

Critical graphs. From these beginnings arose the use of linear algebra to produce results in graph theory, sometimes in surprising ways. One such example was given by the Hungarian mathematician László Lovász in 1973.[4] Recall from Chapter 2 that, for $k > 1$, a k-chromatic graph is k-*critical* if the removal of any vertex or edge lowers the chromatic number. The 3-critical graphs are the odd cycles, and so the maximum number α of independent vertices in a 3-critical graph with n vertices is $\frac{1}{2}(n - 1)$. For 4-critical graphs we have $\alpha \geq \frac{1}{4}n$, but the possible existence of an upper bound of the form cn was answered in the negative by William Brown and John Moon in 1969,[5] who constructed 4-critical graphs with $\alpha \geq n - c_1\sqrt{n}$. Lovász then showed that there is a constant c_2 for which $\alpha \leq n - c_2\sqrt{n}$, so that Brown and Moon's construction is essentially best possible.

Lovász's proof is as follows: If G is a 4-critical graph on n vertices with an independent set S of α vertices with $\alpha \geq n/4$, let each vertex of S be split into vertices all of degree 3, keeping 4-criticality; this results in replacing S by an independent set S' of vertices, all of degree 3, and leaving $G - S$ unchanged. Now form a matrix $\mathbf{M}$ whose rows correspond to pairs of vertices from $G - S$ and whose columns correspond to the vertices of S', where an entry is 1 if the row pair is among the three neighbors of the column vertex, and 0 otherwise. An argument using 4-criticality shows that the columns of $\mathbf{M}$ are linearly independent over the field GF(2) of two elements. But because the column rank of any matrix equals the row rank, $\mathbf{M}$ has at least $|S'|$ rows. Because $(n - \alpha)^2 \geq |S'| \geq |S| = \alpha \geq n/4$, the desired result follows. Thus, an elementary result on matrices (row rank = column rank) leads to a rather deep result on graphs for which no other proof is known.

4.1.2 Eigenvalues of adjacency matrices.

We now move to the 1940s, when a number of chemists and mathematicians began to investigate the eigenvalues of the so-called *Hückel matrix* of a polycyclic hydrocarbon constructed from benzene 6-rings (see Figure 4.3). Suitably scaled, this is essentially the adjacency matrix of the bipartite graph whose vertices are the carbon atoms and whose edges are the carbon–carbon bonds (see Figure 4.4). The energy levels of the molecule correspond to the eigenvalues of the adjacency matrix, and the wave functions correspond to the eigenvectors.

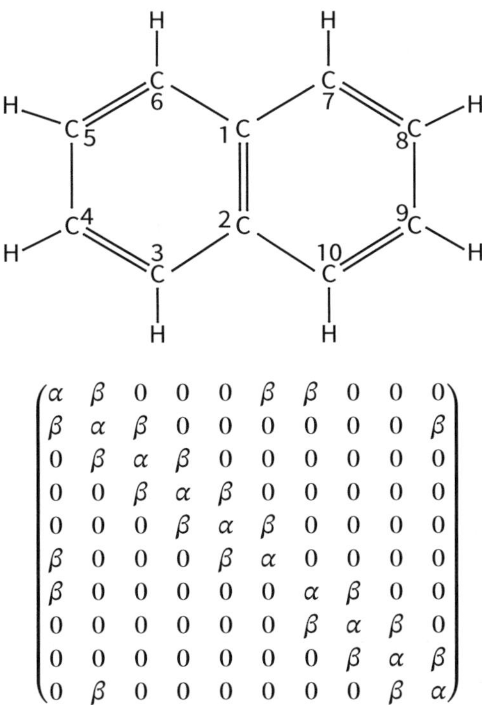

$$\begin{pmatrix}
\alpha & \beta & 0 & 0 & 0 & \beta & \beta & 0 & 0 & 0 \\
\beta & \alpha & \beta & 0 & 0 & 0 & 0 & 0 & 0 & \beta \\
0 & \beta & \alpha & \beta & 0 & 0 & 0 & 0 & 0 & 0 \\
0 & 0 & \beta & \alpha & \beta & 0 & 0 & 0 & 0 & 0 \\
0 & 0 & 0 & \beta & \alpha & \beta & 0 & 0 & 0 & 0 \\
\beta & 0 & 0 & 0 & \beta & \alpha & 0 & 0 & 0 & 0 \\
\beta & 0 & 0 & 0 & 0 & 0 & \alpha & \beta & 0 & 0 \\
0 & 0 & 0 & 0 & 0 & 0 & \beta & \alpha & \beta & 0 \\
0 & 0 & 0 & 0 & 0 & 0 & 0 & \beta & \alpha & \beta \\
0 & \beta & 0 & 0 & 0 & 0 & 0 & 0 & \beta & \alpha
\end{pmatrix}$$

Figure 4.3. A benzene molecule and its associated Hückel matrix

The adjacency matrix $\mathbf{A}$ of a graph G (possibly with loops and multiple edges) is a symmetric matrix which can be diagonalized. Its eigenvalues λ are the solutions of the characteristic equation $\det(\mathbf{A} - \lambda\mathbf{I}) = 0$, and these bear remarkable connections with some of the parameters of G. For example, if G is a simple connected graph with n vertices, m edges, and t triangles (3-cycles), then the eigenvalues $\lambda_1, \lambda_2, \ldots, \lambda_n$ of its adjacency matrix satisfy

$$\sum_i \lambda_i = 0, \sum_i \lambda_i^2 = 2m, \text{ and } \sum_i \lambda_i^3 = 6t.$$

In terms of the characteristic equation

$$\lambda^n + a_1\lambda^{n-1} + a_2\lambda^{n-2} + \ldots + a_{n-1}\lambda + a_n = 0,$$

this means that $a_1 = 0, a_2 = -m$, and $a_3 = -2t$.

The first mathematical papers on the eigenvalues of a graph were published in Russian by L. M. Lichtenbaum in 1956,[6] and in German by Lothar Collatz and Ulrich Sinogowitz in 1957.[7] In their paper, Collatz and Sinogowitz began by interpreting the first five coefficients a_i of the characteristic equation in terms of properties of the graph; these results were later extended by Horst Sachs,[8] who proved that, for each k,

$$a_k = \sum_H (-1)^{\alpha(H)} 2^{\beta(H)},$$

where the summation extends over all k-vertex subgraphs H consisting of $\alpha(H)$ connected components, each of which is a single edge or one of $\beta(H)$ cycles.

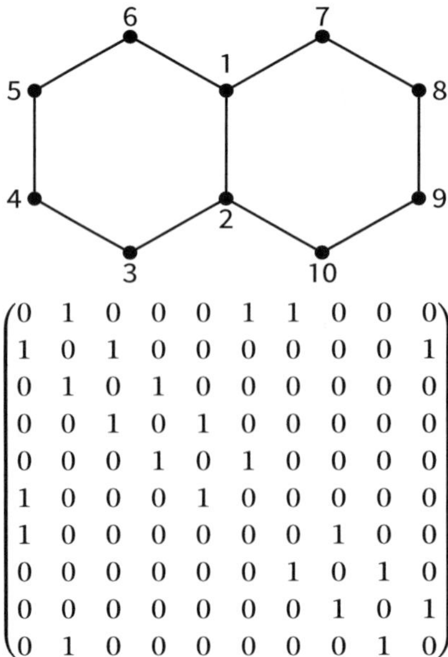

$$
\begin{pmatrix}
0 & 1 & 0 & 0 & 0 & 1 & 1 & 0 & 0 & 0 \\
1 & 0 & 1 & 0 & 0 & 0 & 0 & 0 & 0 & 1 \\
0 & 1 & 0 & 1 & 0 & 0 & 0 & 0 & 0 & 0 \\
0 & 0 & 1 & 0 & 1 & 0 & 0 & 0 & 0 & 0 \\
0 & 0 & 0 & 1 & 0 & 1 & 0 & 0 & 0 & 0 \\
1 & 0 & 0 & 0 & 1 & 0 & 0 & 0 & 0 & 0 \\
1 & 0 & 0 & 0 & 0 & 0 & 0 & 1 & 0 & 0 \\
0 & 0 & 0 & 0 & 0 & 0 & 1 & 0 & 1 & 0 \\
0 & 0 & 0 & 0 & 0 & 0 & 0 & 1 & 0 & 1 \\
0 & 1 & 0 & 0 & 0 & 0 & 0 & 0 & 1 & 0
\end{pmatrix}
$$

Figure 4.4. The associated bipartite graph and its
adjacency matrix

Collatz and Sinogowitz then considered various special types of connected graphs, investigating in particular the eigenvalues of bipartite graphs, graphs with a zero eigenvalue, and graphs with symmetries, and listing explicitly the eigenvalues of complete graphs, path graphs, and the graph of the icosahedron.

Their paper continued with tables of the characteristic polynomials of all the simple connected graphs with up to 5 vertices, the trees with 6, 7, and 8 vertices, and a further 42 graphs with interesting properties, such as the five Platonic graphs and the Petersen graph. As examples, we note that

- the eigenvalues of the complete graph K_n are $n - 1$ (once) and -1 ($n - 1$ times);

- the eigenvalues of the complete bipartite graph $K_{r,s}$ are $\sqrt{rs}$, $-\sqrt{rs}$ and 0 ($r + s - 2$ times);

- the eigenvalues of the cycle graph C_n are $2\cos(2\pi k/n)$, for $k = 0, 1, 2, \dots, n - 1$;

- the eigenvalues of the Petersen graph are 3 (once), 1 (five times), and -2 (four times).

For bipartite graphs, all the eigenvalues are "paired"—that is, if λ is an eigenvalue with multiplicity k, then $-\lambda$ is also an eigenvalue with the same multiplicity. This was proved by mathematicians in the 1950s, but such results were already known to chemists: the *Coulson–Rushbrooke pairing theorem* had asserted the corresponding result for the energy levels of polycyclic molecules in 1940.[9]

The final section of the Collatz and Sinogowitz paper contains results on the largest eigenvalue of a connected graph. It follows from earlier results of Oskar Perron and Georg Frobenius[10] on matrices with nonnegative entries that the largest eigenvalue r of the adjacency matrix of a connected graph is a simple root of the characteristic equation, and is the only eigenvalue that has a corresponding eigenvector with positive coordinates. Moreover, all other eigenvalues lie between $-r$ and r. For a k-regular graph, k is an eigenvalue with $(1, 1, \ldots, 1)$ as a corresponding eigenvector.

We have just considered the eigenvalues of particular types of connected graph, but in recent years the problem has been turned around:

Given certain restrictions on the eigenvalues, what can be deduced about the corresponding graphs?

For example, if the largest eigenvalue is r, and if $-r$ is also an eigenvalue (that is, the largest eigenvalue is paired), then we can deduce that the graph must be bipartite.

We conclude this section with two further results of this type, corresponding to the cases when the largest eigenvalue is 2 and when the smallest eigenvalue r is -2: each of these conditions imposes severe restrictions on the associated graphs. In the first case, John H. Smith[11] showed that if G is a connected graph with $r = 2$, then it is either a cycle or one of the trees in Figure 4.5. In the second case, if the graph G is regular and connected with smallest eigenvalue -2, then it must be one of these three types:

- a line graph (in which the vertices correspond to the edges of a graph H, and two vertices are adjacent whenever the corresponding edges of H meet);

- a d-dimensional octahedron (for some d);

- a graph arising from a set of vectors inclined at angles of 90° and 60° in a "root system" associated with the Dynkin diagram E_8 (see the paper by A. J. Hoffman and D. K. Ray-Chaudhuri and the one by P. J. Cameron et al.).[12]

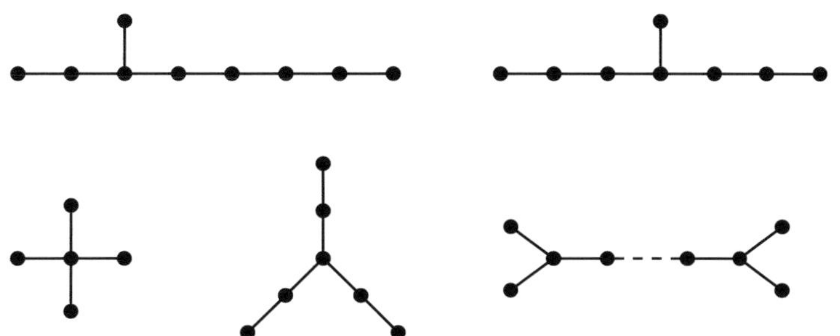

Figure 4.5. The trees with largest eigenvalue 2

The characteristic polynomial that we have just discussed and the chromatic polynomial that we met in Chapter 2 are not the only polynomials whose algebraic properties illuminate areas of graph theory. Others are the Potts model partition function, used to study phase transitions in physics, and the Tutte polynomial. Further information about these polynomials and the connections among them can be found in the survey by Bill Jackson and the book by Dominic Welsh.[13]

4.1.3 Eigenvalues and expanders. We have just given examples to show how restrictions on the eigenvalues can lead to consequences for the corresponding graphs. In recent years, the second-largest eigenvalue of a graph has come to play important roles in a variety of topics, from computer science and complexity theory to network theory and cryptography.

Of particular interest are those connected graphs whose second eigenvalue is much smaller than the largest one; the extreme case is that of the complete graph K_n, whose largest eigenvalue is $n-1$ and whose other eigenvalues are all -1. We seek graphs with small maximum degrees and wide gaps between the largest two eigenvalues. Examples of these include the "Ramanujan graphs", which are regular of degree k and where the second eigenvalue does not exceed

$$2\sqrt{k-1};$$

examples of Ramanujan graphs include complete graphs, the Petersen graph, and the graph of the icosahedron. These graphs are related to Ramanujan's conjectures in number theory (see A. Lubotsky, R. Phillips, and P. Sarnak, and M. Ram Murty).[14]

Expander graphs are connected graphs that do not have too many edges but have high connectivity, subject to the requirement that to disconnect them one has to delete many edges. The *edge-expansion ratio* of a graph G is

$$h(G) = \frac{\min_S\{\text{number of edges joining vertices in } S \text{ with vertices not in } S\}}{|S|},$$

where the minimum is taken over all sets S with at most half of the vertices of G. Good expanders have a large edge-expansion ratio, and they tend to have a large gap between the largest two eigenvalues; for example, Ramanujan graphs tend to be good expanders, as do random graphs (see Chapter 7). In 1986, several authors proved versions of the result that, if G is a regular graph, then

$$(\lambda_1 - \lambda_2)/2 \leq h(G) \leq \sqrt{2\lambda_1(\lambda_1 - \lambda_2)},$$

where λ_1 is the largest eigenvalue and λ_2 is the second largest; for details, see the excellent survey by S. Hoory, N. Lineal, and A. Wigderson.[15]

There are several different types of expanders, depending on the problem in hand; for example, an (n, d, c)-expander is a bipartite graph with n vertices in each of the two partite sets, 'inputs' and 'outputs', and with maximum degree d, in which, for some fixed number c and for each set S of at most $n/2$ inputs, there are at least $(1 + c(1-|S|/n))|S|$ neighboring outputs. In 1986, Noga Alon proved that a regular bipartite graph is an expander in this sense if and only if the largest and second largest eigenvalues are well separated.[16]

Also of interest is the related idea of a *superconcentrator*, introduced by L. G. Valiant[17] in 1976. This is a digraph without any directed cycles in which there are two sets of n vertices (inputs and outputs) with the property that, for any set S of inputs and any set T of outputs with $|S| = |T|$, there are $|S|$ vertex-disjoint paths from S to T. In 1984, Noga Alon and V. D. Milman surveyed some results on expanders and superconcentrators, outlined some connections between the eigenvalues of a graph and its expansion properties, and showed how to construct expanders and superconcentrators with particular properties.[18]

4.2 Graphs and groups

Some graphs are more symmetrical than others; for example, of the three graphs in Figure 4.6, the complete graph K_5 seems highly symmetrical, the cycle graph C_5 with one additional edge has just one nontrivial symmetry (corresponding to a reflection), and the 6-vertex graph has none.

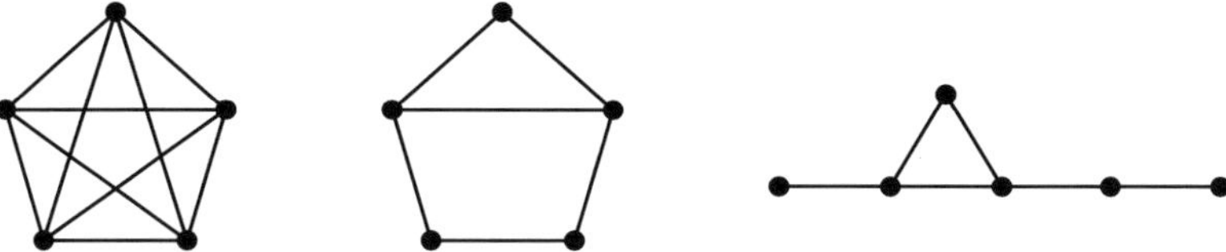

Figure 4.6. Graphs with varying degrees of symmetry

4.2.1 Automorphism groups.

We can formalize these intuitive ideas as follows. An *automorphism* of a graph G is a permutation φ of the vertex-set of G with the property that $\varphi(v)\varphi(w)$ is an edge of G if and only if vw is. There is a similar definition for directed graphs, where an automorphism must also preserve the directions of the arcs.

The automorphisms of a graph G form a group Aut G under composition, called the *automorphism group* of G; for example, the graph K_5 has $5! = 120$ automorphisms, corresponding to the 120 possible permutations of its vertices, and its automorphism group is the symmetric group $\mathbb{S}_5$. The cycle graph C_5 has ten automorphisms, corresponding to the rotations and reflections of a regular pentagon, and its automorphism group is the dihedral group $\mathbb{D}_5$ of order 10, but the middle graph above has only a reflection and the identity automorphism, so its automorphism group is the group of order 2. The 6-vertex graph in the figure has only the identity automorphism, so its automorphism group is the trivial group; it is the smallest graph with this property.

For planar graphs with at least two vertices, a surprising result of V. A. Aksionov et al.[19] is that for such a graph the deletion of at most five edges results in a graph with nontrivial automorphism group; this result is best possible since the statement is false if "five" is replaced by "four". In general, most graphs have no symmetries and therefore have the trivial automorphism group; this means, in particular, that each vertex can be distinguished from each of the others. This is useful when one wishes to check pairs of graphs for possible isomorphism.

We can turn the situation around and ask whether every finite group is the automorphism group of some graph, a question posed by Dénes König in his celebrated textbook of 1936 (see Chapter 1). The answer was given in 1938 in a classic paper by the German–Chilean mathematician Roberto Frucht, who showed how to construct infinitely many graphs arising from any given group.[20] But we can say more. In a subsequent paper,[21] Frucht proved that every finite group is the automorphism group of some cubic graph, and similar statements can be made if we replace "cubic" by "bipartite", "Hamiltonian", "k-regular", "k-connected", or "k-chromatic" (for any $k \geq 2$). However, as George Pólya discovered, not every finite group is the automorphism group of a tree.

For directed graphs the situation is similar since every finite group is also the automorphism group of some digraph. But the statement does not hold for certain specific classes of digraphs, such as tournaments. A *tournament* is an oriented complete graph—that is, a complete graph in which one direction has been assigned to each edge (see Figure 4.7). In 1964, John Moon explained why no tournament can have an automorphism group of even order, and proved that every group with an odd number of elements is the automorphism group of infinitely many tournaments.[22] Furthermore, these tournaments can be taken to be strongly connected—that is, there is a directed path from any vertex to any other.

Over the years there has been much work linking different types of groups with the properties of corresponding graphs. In particular, in 1980 Peter Cameron posed the following question:[23]

Given that a large graph admits a group of automorphisms isomorphic to the abstract group G, what is the probability that G is its full automorphism group?

4.2.2 Cayley graphs. Another link between groups and graphs was introduced by Arthur Cayley[24] in the first volume of the *American Journal of Mathematics*, founded in 1878 by his friend J. J. Sylvester (see Chapter 1). Given a finite permutation group Γ, let H be a subset of Γ that does not contain the identity element and is closed under the taking of inverses (so that if h is in H, then so is h^{-1}). The *Cayley graph* $\mathrm{Cay}(\Gamma, H)$ is the graph whose vertices are the elements of Γ and whose edges are the pairs of the form $\{g, gh\}$, where g is in Γ and h is in H; for example, if Γ is the cyclic group $\mathbb{Z}_n$ of order n, and H consists of a generator of $\mathbb{Z}_n$ and its inverse, then $\mathrm{Cay}(\Gamma, H)$ is the cycle graph C_n.

It is natural to ask when a given graph is a Cayley graph. This question was answered by Gert Sabidussi in 1958, in a paper that has proved to be fundamental in most later discussions on the subject.[25] It focuses on the properties of strongly fixed-point-free graphs, which are graphs with nontrivial automorphism group for which $\varphi(v) \neq v$ for every nontrivial automorphism φ and every vertex v in F.

4.2.3 Graphs with symmetry. Many of the most interesting graphs have some degree of symmetry—these are graphs (often regular) whose automorphism group is nontrivial. In this connection, much attention has been paid to Moore graphs, strongly regular graphs, and transitivity.

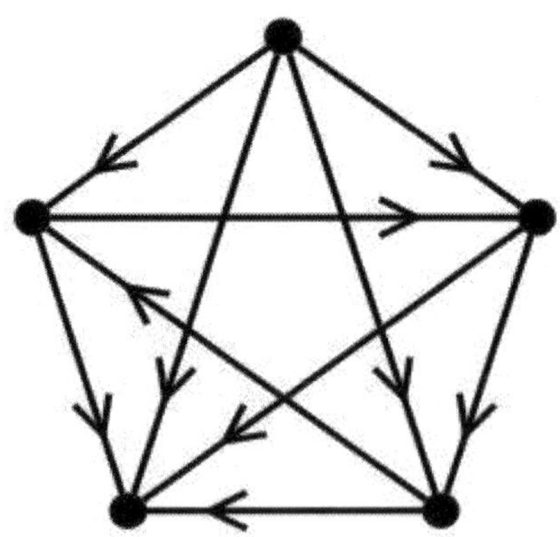

Figure 4.7. A tournament

The *distance* between two vertices of a graph G is the number of edges in a shortest path connecting them, and the *diameter* of G is the maximum of the distances between two vertices; for example, the 5-cycle and the Petersen graph both have diameter 2, while the 7-cycle has diameter 3.

Moore graphs. Let G be a d-regular graph with n vertices, let v be a specified vertex of G, and let n_i be the number of vertices at distance i from v. Then $n_0 = 1$ and $n_i \leq d(d-1)^{i-1}$ for each $i \geq 1$, and so

$$n = \sum_{i \geq 1} n_i \leq 1 + d \sum_{i \geq 1} (d-1)^{i-1}.$$

If equality holds here, then G is a *Moore graph*, named after the American mathematician and computer scientist Edward F. Moore. Moore graphs with diameter 2 and degree d have $d^2 + 1$ vertices, and those with diameter 3 have $d^3 - d^2 + d + 1$ vertices.

In 1960, Alan Hoffman and Robert Singleton determined the Moore graphs with diameters 2 and 3, obtaining the following list:[26]

Diameter 2

- the 5-cycle (degree 2);

- the Petersen graph (degree 3);

- the *Hoffman–Singleton graph* with 50 vertices and degree 7, shown in Figure 4.8;

- and possibly a graph with 3250 vertices and degree 57, whose existence is still undetermined.

Diameter 3

- the 7-cycle (degree 2).

In 1973, E. Bannai and T. Ito, and Mark Damerell,[27] proved independently that there are no Moore graphs with diameter greater than 3.

Strongly regular graphs. Strongly regular graphs were introduced by R. C. Bose in a classic paper of 1963.[28] A *strongly regular graph* with parameters (n, k, λ, μ) is a graph with n vertices that is regular of degree k, and in which any two adjacent vertices have exactly λ common neighbors and any two nonadjacent vertices have exactly μ common neighbors; these parameters are linked by the equation $k(k - \lambda - 1) = (n - k - 1)\mu$. Every connected strongly regular graph has diameter 2 and has just three different eigenvalues: the degree k and the roots of the quadratic equation

$$x^2 = k + \lambda x - \mu(1 + x);$$

for example, the Petersen graph is a strongly regular graph with parameters $(10, 3, 0, 1)$ and distinct eigenvalues $3, 1,$ and -2.

Transitive graphs. Another much-studied property of regular graphs is transitivity. A graph is *vertex-transitive* if there is an automorphism that sends any given vertex to any other; such graphs must be regular, and examples include the graphs K_n and C_n, for $n \geq 3$. Similarly, a graph is *edge-transitive* if there is an automorphism that sends any given edge to any other; such graphs need not be regular, and examples include

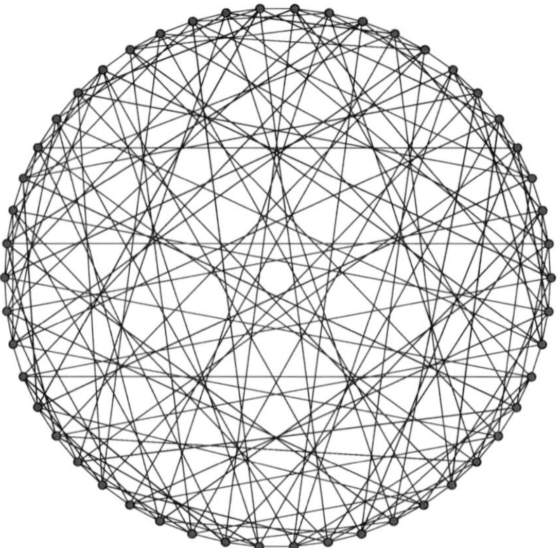

Figure 4.8. The Hoffman–Singleton graph with 50
vertices

the complete bipartite graph $K_{r,s}$ when $r \neq s$. The Petersen graph has both of these
properties.

We conclude with a much stronger symmetry condition. Generalizing the idea of
edge-transitivity, where there is an automorphism that sends any two adjacent vertices
to any other such pair, we define a graph to be *distance-transitive* if, whenever the distance between two vertices v and w equals that between the vertices x and y, there is
an automorphism of the graph that sends v to x and w to y. Examples of distance-
transitive graphs are, for each $n \geq 3$, the complete graph K_n, the regular complete
bipartite graph $K_{n,n}$, the cycle graph C_n, the d-dimensional cube Q_d, and the Petersen
graph. In an attractive paper of 1971, Norman Biggs and Derek Smith[29] determined all
distance-transitive graphs that are regular of degree 3.

4.3 Conclusion

Over the past century, the links between graph theory and algebra have grown ever
stronger. With the development of the former from a collection of largely unconnected
results studied by a handful of enthusiasts to a fully accepted area of mathematics with
researchers from around the globe, and with the spectacular advances in the latter,
with such achievements as the Feit–Thompson theorem on solvable groups and the
classification of finite simple groups, the two subjects have been seen to have a great
deal in common, with each contributing to the other.

It is likely that part of the reason for these developments has been the rise of the
computer, whose use has led to the investigation of increasingly larger graphs and
groups of symmetries. It is perhaps not surprising that two of the earliest (and most no-
torious) uses of computers in mathematics were the classification of the finite simple
groups and the proof of the four-color theorem.

Further Reading

For further details on the topics of this chapter see the following multi-authored volume:

> Lowell W. Beineke and Robin J. Wilson (eds.), *Topics in Algebraic Graph Theory*, Cambridge Univ. Press (2004).

> See, in particular, these survey chapters: "Eigenvalues of graphs" by Michael Doob, "Automorphisms of graphs" and "Strongly regular graphs" by Peter Cameron, and "Cayley graphs" by Brian Alspach.

For further information on graph theory and linear algebra, see

> N. L. Biggs, *Algebraic Graph Theory*, Cambridge Univ. Press (1994);

> D. Cvetković, M. Doob, and H. Sachs, *Spectra of Graphs*, VEB (1980); 3rd edn., J. A. Barth (1997);

> C. Godsil and G. Royle, *Algebraic Graph Theory*, Springer Graduate Texts in Math. 207, Springer (2001).

For a gentle introduction to groups and graphs, see

> E. Grossman and W. Magnus, *Groups and their Graphs*, MAA (1964).

Further information may be found in

> P. J. Cameron, *Permutation Groups*, Cambridge Univ. Press (1999);

> W. Dicks and M. J. Dunwoody, *Groups Acting on Graphs*, Cambridge Univ. Press (1989).

Notes and References

[1] M. Dehn and P. Heegaard, Analysis Situs, *Encyklopädie der Mathematischen Wissenschaften* IIIAB3 (1907), 153–220. Max Dehn was one of the first to solve one of David Hilbert's celebrated problems from the 1900 International Congress of Mathematicians in Paris; because of his Jewish background, he emigrated in 1939 from Germany, via Denmark and Norway, to the United States and taught in North Carolina. Poul Heegaard was a student of Julius Petersen. In 1898, his doctoral thesis laid the foundation of modern knot theory and contained a counterexample to the version of Poincaré duality that Poincaré had just published. From 1918 to 1941, he taught at the University of Oslo in Norway. For a detailed biography by Ellen S. and Hans J. Munkholm, see Chapter 34 of *History of Topology*, Elsevier (1999).

[2] O. Veblen, An application of modular equations in analysis situs, *Ann. of Math.* (2) 14 (1912/13), 86–94. Widely remembered for his contributions to geometry, Veblen published this paper in the same year that George Birkhoff introduced chromatic polynomials (see Chapter 2).

[3] O. Veblen, *Analysis Situs*, Amer. Math. Soc. Colloq. Lectures, 1916, New York (1922).

[4] L. Lovász, Topological and algebraic methods in graph theory, *Graph Theory and Related Topics* (eds. J. A. Bondy and U. S. R. Murty), Academic Press (1979), 1–14. In 2021, László Lovász and Avi Wigderson (see Note 15) were jointly awarded the Abel Prize for their foundational work in computer science and discrete mathematics.

[5] W. G. Brown and J. W. Moon, Sur les ensembles de sommets indépendants dans les graphes chromatiques minimaux, *Canad. J. Math.* 21 (1969), 274–278. The two authors, both English-speaking, wrote their paper in French and submitted it in July 1967, in celebration of the centennial of the Dominion of Canada during that month.

[6] L. M. Lichtenbaum, Eigenvalues of a simple graph (Russian), *Trudy 3-vo Vsesoyuznovo Mat. Syezda*, 1 (1956), 135.

[7]L. Collatz and U. Sinogowitz, Spektren endlicher Grafen, *Abh. Math. Sem. Univ. Hamburg* 21 (1957), 63–77. Sinogowitz was killed in an air raid over Darmstadt, Germany, in 1944, but had already made substantial contributions to this paper.

[8]H. Sachs, Beziehungen zwischen den in einem Graphen enthaltenen Kreisen und seinem charakteristischen Polynom, *Publ. Math. Debrecen* 11 (1964), 119–134. Sachs was the leading graph theorist in East Germany; this result first appeared in his thesis at Halle University in 1963.

[9]C. A. Coulson and G. S. Rushbrooke, Note on the method of molecular orbitals, *Proc. Cambridge Philos. Soc.* 36 (1940), 193–200. After teaching at the Universities of St Andrews and Cambridge, Charles Coulson held professorial posts in physics in London and in applied mathematics and theoretical chemistry in Oxford.

[10]O. Perron, Zur Theorie der Matrizen, *Math. Ann.* 64 (1907), 248–263, and G. Frobenius, Über Matrizen aus nicht negativen Elementen, *Sitzber. Akad. Wiss. Berlin* (1912), 456–477. Perron obtained the result for matrices with positive entries, and Frobenius extended it to nonnegative entries.

[11]J. H. Smith, Some properties of the spectrum of a graph, *Combinatorial Structures and their Applications* (eds. R. K. Guy *et al.*), Gordon and Breach (1970), 403–406.

[12]A. J. Hoffman and D. K. Ray-Chaudhuri, On the line graph of a symmetric balanced incomplete block design, *Trans. Amer. Math. Soc.* 116 (1965), 238–252 and P. J. Cameron, J. M. Goethals, J. J. Seidel, and E. E. Shult, Line graphs, root systems, and elliptic geometry, *J. Algebra* 43 (1976), 305–327.

[13]B. Jackson, Chromatic polynomials, *Topics in Chromatic Graph Theory* (eds. L. W. Beineke and R. J. Wilson), Cambridge Univ. Press (2015), 56–72 and D. J. A. Welsh, *Complexity, Knots, Colouring and Counting*, London Math. Soc. Lecture Note Series 186, Cambridge Univ. Press (1993).

[14]A. Lubotsky, R. Phillips, and P. Sarnak, Explicit expanders and the Ramanujan conjectures, *Proc. 18th Ann. ACM Symp. Theory of Computing* (1986), 240–246 and M. Ram Murty, Ramanujan graphs, *J. Ramanujan Math. Soc.* 18 (2003), 1–20.

[15]S. Hoory, N. Lineal, and A. Wigderson, Expander graphs and their applications, *Bull. Amer. Math. Soc.* 43 (2006), 439–561.

[16]N. Alon, Eigenvalues and expanders, *Combinatorica* 6 (1986), 83–96.

[17]L. G. Valiant, Graph theoretic properties in computational complexity, *J. Comput. Syst. Sci.* 13 (1976), 278–285.

[18]N. Alon and V. D. Milman, Eigenvalues, expanders and superconcentrators, *Proc. 25th Symp. Foundations of Computer Science*, IEEE (1984), 320–322.

[19]V. A. Aksionov, O. V. Borodin, L. S. Melnikov, G. Sabidussi, M. Stiebitz, and B. Toft, Deeply asymmetric planar graphs, *J. Combin. Theory* (B) 95 (2005), 68–78. This paper constructs planar nonsymmetric graphs from which the removal of up to four edges never results in symmetry.

[20]R. Frucht, Herstellung von Graphen mit vorgegebener abstrakter Gruppe, *Compos. Math.* 6 (1939), 239–250.

[21]R. Frucht, Graphs of degree 3 with given abstract group, *Canad. J. Math.* 1 (1949), 365–378; see also F. Harary, Homage to Roberto Frucht, *J. Graph Theory* 6 (1982), 97–99.

[22]J. W. Moon, Tournaments with a given automorphism group, *Canad. J. Math.* 16 (1964), 485–489. Moon also published a monograph, *Topics on Tournaments*, Holt, Rinehart and Winston (1968).

[23]P. J. Cameron, On graphs with given automorphism group, *European J. Combin.* 1 (1980), 91–96.

[24]A. Cayley, Desiderata and suggestions: No. 2. The theory of groups: graphical representation, *Amer. J. Math.* 1 (1878), 174–176 = *Coll. Math. Papers* 10, 403–405.

[25]G. Sabidussi, On a class of fixed-point-free graphs, *Proc. Amer. Math. Soc.* 9 (1958), 800–804.

[26]A. J. Hoffman and R. R. Singleton, On Moore graphs with diameters 2 and 3, *IBM J. Res. Develop.* 4 (1960), 497–504.

[27]E. Bannai and T. Ito, On Moore graphs, *J. Fac. Sci. Univ. Tokyo* (A) 20 (1973), 191–208 and R. M. Damerell, On Moore graphs, *Proc. Cambridge Philos. Soc.* 74 (1973), 227–236.

[28] R. C. Bose, Strongly regular graphs, partial geometries, and partially balanced designs, *Pacific J. Math.* 13 (1963), 389–419.

[29] N. L. Biggs and D. H. Smith, On trivalent graphs, *Bull. London Math. Soc.* 3 (1971), 155–158.

5

Cycles, Factorizations, and Matchings

Cycles have played an important role in the development of graph theory and have been studied intensively. As we mentioned in Chapter 2, Dénes König proved in 1916 that a graph is bipartite if and only if it has no cycles of odd length, while in Chapter 1 we saw that William Rowan Hamilton investigated cycles that pass through all the vertices of a dodecahedron. As we shall discover in Section 5.1, these Hamiltonian cycles also play a role in the four-color problem, because the regions of a map containing such a cycle can be colored with four colors. There are also numerous papers that relate cycle lengths to other properties of graphs, but characterizing Hamiltonian graphs in general has proved elusive.

The remainder of this chapter introduces the related topics of factorizations and matchings. A *spanning subgraph* of a graph G is a subgraph which includes every vertex of G, and a *2-factor* in G is a spanning subgraph that is regular of degree 2; if G is Hamiltonian, then a Hamiltonian cycle is an example of a 2-factor in G. Similarly, a *1-factor* is a spanning subgraph that is regular of degree 1, and a *k-factor* is a spanning subgraph that is regular of degree k.

Characterizing the graphs with a 2-factor was first considered in the late 1880s by J. J. Sylvester in connection with the factorization of polynomials (as we shall explain in Section 5.2). An intensive correspondence between Sylvester and the Danish mathematician Julius Petersen then led in 1891 to Petersen's famous paper on regular graphs and their 1- and 2-factors. Possibly the earliest paper to consider problems motivated by graph theory itself, rather than from applications, it led to major developments in the 20th century.

The general problem of determining when a graph has a 1-factor, also called a *perfect* or *complete matching*, was solved much later by W. T. (Bill) Tutte, whom we met in Chapter 2. For bipartite graphs, a solution had been given in 1916 by König and generalized by him and by Jenő Egerváry in 1931 (see König's theorem in Section 5.2.2 and about Egerváry's contribution in Chapter 8). Tutte's theorem is a "good theorem", in the sense that the nonexistence of a 1-factor has an explanation that can be checked efficiently; this use of the word "good" was coined by Jack Edmonds, who in 1965 extended Tutte's theorem by providing a good (that is, polynomial-time) algorithm for the 1-factor problem.

5.1 Hamiltonian graphs

In Chapter 1, we saw how Hamiltonian graphs arose from investigations into cycles on polyhedra by Hamilton and T. P. Kirkman, and from Euler's solution of the knight's tour problem. In each case, the aim was to find a closed cycle that passes through every vertex of the associated graph before returning to the start.

In 1880, P. G. Tait conjectured that such cycles can be found in any cubic polyhedron—that is, one with exactly three faces meet at each vertex. This would then prove the four-color theorem, because every map can be reduced to one that is cubic, and if such a map has such a Hamiltonian cycle passing through all of its vertices, then we can 4-color its regions by coloring the regions inside the cycle red and blue and those outside the cycle green and yellow.

Kirkman remarked that Tait's conjecture "mocks alike at doubt and proof", meaning that he could not make up his mind as to its truth. However, Tait's conjecture is false—there is no Hamiltonian cycle in the cubic graph in Figure 5.1 presented by Tutte in 1946.[1] Systematic methods for finding such counterexamples were later given by V. Kozyrev of Moscow and E. J. Grinberg of Riga, as recorded by Horst Sach.[2]

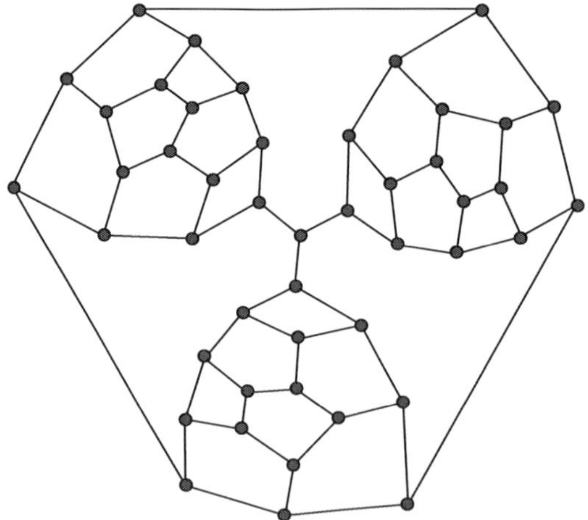

Figure 5.1. Tutte's non-Hamiltonian polyhedral graph

Around 1890, various authors looked at cycles passing through every vertex of a graph, and further investigations took place in 1917 by A. Kowalewski and in 1926 by André Sainte-Laguë. But it was not until 1936 that Hamiltonian graphs were discussed in greater detail, in König's classic text (see Chapter 1), and that the term "Hamiltonian" became firmly established.[3]

Directed graphs can also have Hamiltonian cycles. As we saw in Chapter 4, a *tournament* is a complete directed graph, and among the results in König's book is a theorem by his Hungarian countryman László Rédei, who proved in 1934 that every tournament has a Hamiltonian path (a directed path through all the vertices), and that the number of such paths is odd.[4] Rédei actually proved a stronger version of this result in his paper, but it was only in 1959 that it was established by Paul Camion[5] that every strongly connected tournament (see Chapter 4) has a directed Hamiltonian cycle—in fact, every such tournament has cycles of all possible lengths.

In 1951, Gabriel Dirac completed his Ph.D. degree at the University of London. His thesis dealt with the coloring of graphs and, as we saw in Chapter 2, it introduced and studied critical graphs. It also contained theorems relating lengths of paths and cycles to vertex-degrees, and in particular, he proved the following celebrated result.

Theorem 5.1 (Dirac's theorem). *If G is a graph with n vertices ($n \geq 3$) in which the degree of each vertex is at least n/2, then it is Hamiltonian.*

If G is a critical graph with chromatic number k, then every vertex has degree at least $k - 1$, and it follows that if G has at most $2k - 2$ vertices, then it is Hamiltonian. In subsequent papers Dirac published this and other results from his thesis, of which his paper of 1952 is perhaps his most important.[6]

Dirac's theorem on Hamiltonian cycles has been generalized several times, by Oystein Ore, Lajós Posa, Vašek Chvátal, and others. Ore's result, in a one-page paper from 1960, is the following theorem.[7]

Theorem 5.2 (Ore's theorem). *If G is a graph with n vertices in which deg v + deg w $\geq$ n for each pair of nonadjacent vertices v and w, then G is Hamiltonian.*

Ore's theorem received a substantial extension in 1976 by Adrian Bondy and Vašek Chvátal,[8] who defined the *closure* of a graph. The closure operation on a graph G with n vertices is obtained by successively adding edges between pairs of nonadjacent vertices whose degree-sum is at least n, as illustrated in Figure 5.2. For graphs satisfying Dirac's or Ore's condition, the closure is a complete graph, and in every case the procedure always leads to a unique result. The general theorem is the following.

Theorem 5.3 (Bondy and Chvátal's theorem). *A graph is Hamiltonian if and only if its closure is Hamiltonian.*

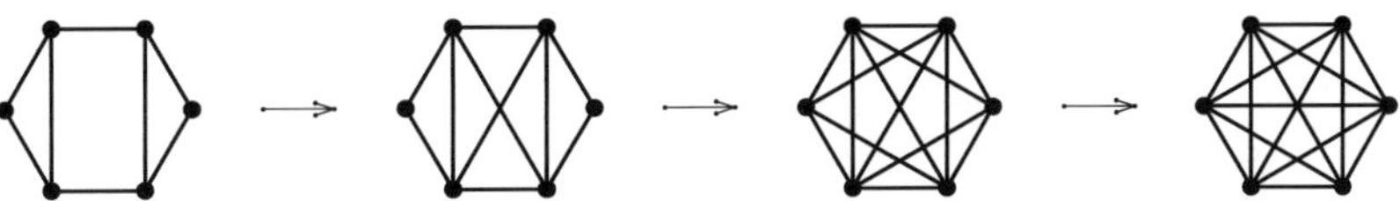

Figure 5.2. Forming the closure of a graph

5.2 Factorizations and matchings

In 1889, J. J. Sylvester reached the age of 75. Suffering from serious eye problems he travelled to Sweden for a possible cure, and on the return journey he visited Copenhagen to meet Julius Petersen and Hieronymus Zeuthen, the two mathematics professors at Copenhagen University. Petersen took Sylvester to the Tivoli Gardens where they discussed the latest mathematical sensation, David Hilbert's proof of the finite basis theorem for binary invariants. Sylvester told Petersen that a much simpler proof, avoiding Hilbert's reliance on Diophantine equations, had been found by his friend Arthur Cayley in London. Sylvester outlined Cayley's argument, but Petersen found it to be faulty. Back in Oxford, Sylvester wrote to Petersen:[9]

> I deem it very fortunate that I had the chance of talking the matter over with you at Tivoli (!) and having my eyes opened to the wonderful fallacy into which so great a Genius and so practised a Veteran as Cayley had allowed himself to be betrayed as if he had been some novice for the first time grappling with an arduous question.

This episode led to an extensive correspondence between Petersen and Sylvester, and also led to a visit by Petersen to Oxford and London in January 1890.[10] Hilbert's proof had been nonconstructive, and Sylvester and Petersen tried to construct the basis in the special case of homogeneous polynomials in n variables $x_1, x_2, \ldots, x_n$ of the form

$$(x_1 - x_2)^{\alpha(1,2)}(x_1 - x_3)^{\alpha(1,3)} \ldots (x_1 - x_n)^{\alpha(1,n)}(x_2 - x_3)^{\alpha(2,3)} \ldots (x_{n-1} - x_n)^{\alpha(n-1,n)},$$

where the exponents $\alpha(i,j)$ are nonnegative integers, and the maximum degrees of the variables are the same. Such polynomials can be factorized into products of basis polynomials of the same type and having no proper homogeneous factors, and for a fixed value of n the number of such basis polynomials is finite.

5.2.1 Factorizations.

In 1878, Sylvester had translated the above algebraic question into graphical language; this appeared in the note in *Nature* in which he introduced the term *graph* and in a paper in the *American Journal of Mathematics* (see Chapter 1).[11] Eleven years later, he wrote to Petersen explaining how to produce a regular graph from the above homogeneous polynomial by taking n vertices, one for each variable, and joining each pair of vertices i and j by $\alpha(i,j)$ edges, with multiple edges allowed. A homogeneous factor of the polynomial then corresponds to a graph factor in the graph, thereby explaining the use of algebraic terminology in graph theory. The basis polynomials then correspond to graphs without factors, and for a given n the number of such basis graphs with n vertices is finite, whereas the number of regular graphs with n vertices is infinite because multiple edges are allowed.

Sylvester continued his letter with this assertion:

> Let G be any graph in which the frequency [degree] of each vertex is 2 or more (I do not require the frequency to be the same for each vertex). Then a graph T of uniform frequency 2 and another graph G' may be found such that G may be produced by the superposition of T with G'.

However, his assertion is not always true. The assertion is that a graph has a 2-factor if all vertices have degree 2 or more, but this is not so: simple counterexamples are the complete bipartite graphs $K_{r,s}$ with $r \neq s$, such as the graph $K_{2,3}$ in Figure 5.3.

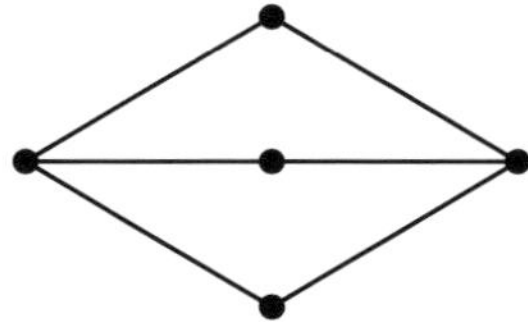

Figure 5.3. A graph of minimum degree 2 with no 2-factor

Sylvester soon realized his mistake and restricted his attention to regular graphs, but he then showed this also to be incorrect by producing a 3-regular graph that has no proper factors (see Figure 5.4). Here, some concepts will be relevant. A *bridge* is an edge whose deletion increases the number of components (it is always by 1), and a *leaf* in a graph is a bridge-free connected component of the graph with a bridge having been removed.

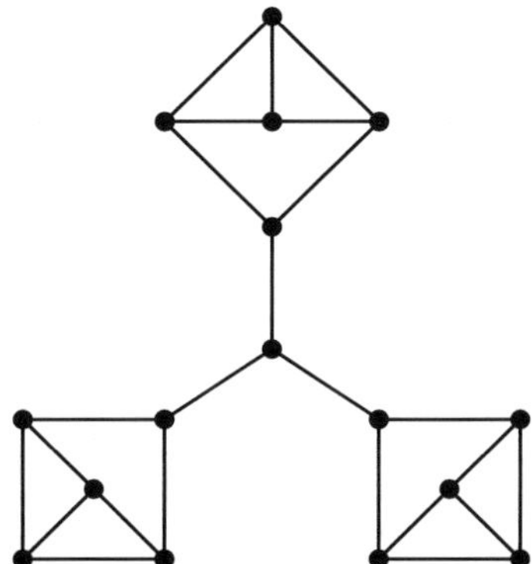

Figure 5.4. Sylvester's 3-regular graph with no proper factors

Petersen eventually took the lead, and he was particularly concerned with regular graphs that can be split into 1-factors and 2-factors. For example, a splitting of the complete graph K_7 into three 2-factors is shown in Figure 5.5.

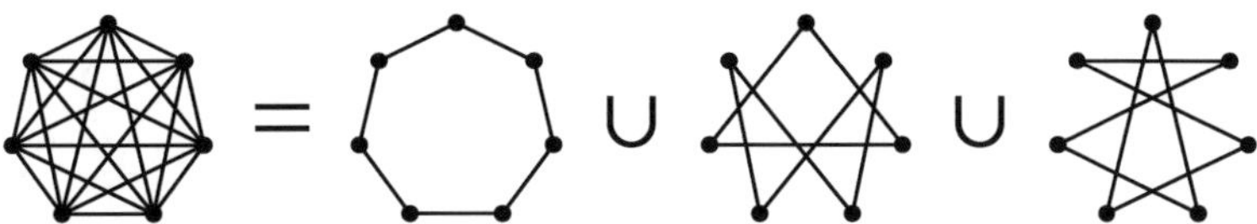

Figure 5.5. The edges of K_7 split into three 2-factors

In 1891, he wrote his celebrated paper,[12] which contained the following fundamental results.

Theorem 5.4 (Petersen's theorem).

(a) *Every 2k-regular graph has a 2-factor, and therefore can be split into 2-factors.*

(b) *Every cubic graph with at most two leaves can be split into a 2-factor and a 1-factor.*

It follows from Petersen's theorem that a 3-regular graph with no 2-factor must have at least three leaves, such as the Sylvester graph in Figure 5.4. To prove part (a) of his theorem, Petersen first reduced the general case to a 4-regular one, and then settled the 4-regular case by producing an Eulerian trail and coloring its edges alternately red and blue. In their correspondence, Sylvester criticized Petersen for making the similar claim that any 6-regular graph splits into two 3-factors; this is false, with the complete graph K_7 as a counterexample.

Sylvester tried in vain to obtain part (a) of Petersen's even-factor theorem by an inductive argument, suggesting various stronger versions to make the induction argument work; in fact, his letters to Petersen included eight such strengthenings, some true and some false—for example, it is false that, for any k edges in a $2k$-regular graph, one can split the graph into k 2-factors, each containing exactly one of the given edges. In his paper, inspired by Sylvester, Petersen paid considerable attention to such separation questions.

Because he had received little input from Sylvester, Petersen published his paper on his own, but it bore marks of the past. With Hilbert's settling of the finite basis theorem, invariant theory was rapidly on the wane. Graph theory still remained a rather undeveloped subject, and it was not until 1936 that König's book brought Petersen's work into prominence, with two chapters devoted to his work. Today we regard Petersen's 1891 paper as pioneering, presenting graph theory seriously on its own merits and with some remarkable theorems.

Later, in 1898, Petersen presented a counterexample to a suggestion that a bridge-free 3-regular graph can be split both into a 1-factor and a 2-factor and into three 1-factors.[13] This is the celebrated *Petersen graph*, which was illustrated in Chapter 1. In Figure 5.6, his drawing is on the left, while the usual drawing (in the middle) seems to have been introduced by C. S. Peirce. A. B. Kempe had previously met the graph while studying Desargues's configuration in geometry;[14] in his drawing (on the right), the vertices represent its ten lines, and the edges correspond to pairs of lines that do not meet at a point of the configuration.

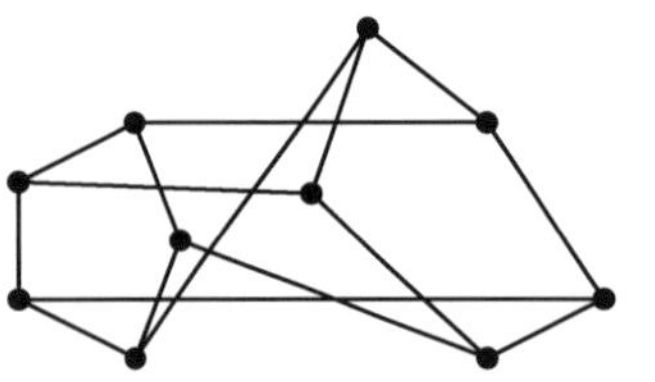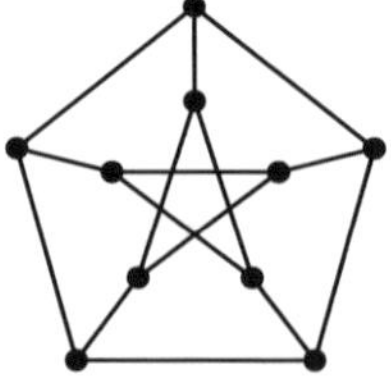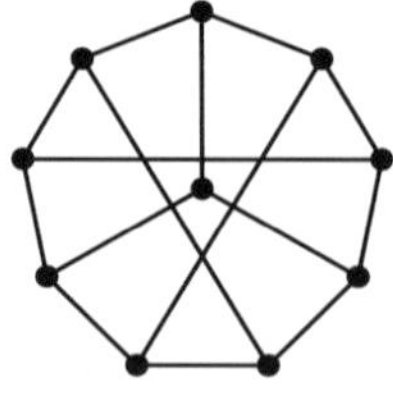

Figure 5.6. Three drawings of the Petersen graph.

5.2.2 Matchings. As we noted in Chapter 2, König proved in 1916 that every k-regular bipartite graph G has a k-edge-coloring, each color of which produces a 1-factor, and so G can be split into k 1-factors. In 1931, he proved a more general theorem that can be considered as a forerunner of the duality theorem of linear programming.[15] It is an example of a "minimax theorem", in which the maximum number of one quantity is equal to the minimum number of another. We state König's theorem in terms of edges, no two of which share a common vertex, called *independent edges*.

Theorem 5.5 (König's theorem). *The maximum number of independent edges in a bipartite graph is equal to the minimum number of vertices that together cover all of the edges.*

König's theorem is related to a new topic that concerns questions of the form: "How connected is a given network?"; one practical application is to the vulnerability of electrical networks. An answer was given in 1927 by Karl Menger,[16] and is another example of a minimax theorem.

Theorem 5.6 (Menger's theorem). *The maximum number of edge-disjoint paths between two vertices of a graph is equal to the minimum number of edges whose removal separates the two vertices into different components.*

Figure 5.7 shows a graph with five edge-disjoint paths between two vertices s and t, and five edges whose removal separates s from t; the maximum number of such edge-disjoint paths and the minimum number of such edges is therefore 5. There is also a version of Menger's theorem concerning the number of internally *vertex-disjoint* paths; here there are three such paths.[17]

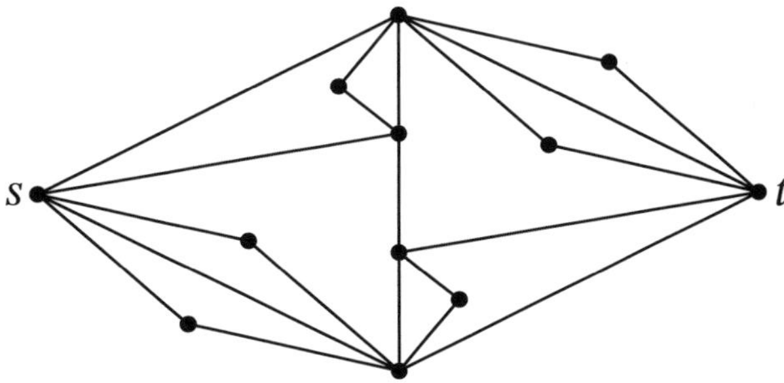

Figure 5.7. An example that illustrates Menger's theorem

Menger's inductive proof was more geometrical than combinatorial. Unfortunately, it was incomplete, as König soon discovered, with his own theorem above as the omitted case.

In another variation of Menger's theorem for directed graphs, there are values on the edges, and these limit the *flow* from s to t. The result is called the *max flow – min cut theorem*, and was first published by L. R. Ford Jr. and D. R. Fulkerson in 1956 (see Chapter 8).[18] In essence, it is the following statement.

Theorem 5.7 (Max flow – min cut theorem). *The maximum amount that can flow from vertex s to vertex t in a given network is equal to the minimum sum of the values on edges that will block the transporting of an amount from s to t.*

A mixed version of Menger's theorem, introduced by Lowell Beineke and Frank Harary, involves the removal of both vertices and edges: An $s - t$ *connectivity pair* (a, b) is a set of a vertices and b edges whose removal separates s from t, but where the removal of fewer vertices or edges leaves s and t in the same component. The "mixed-connectivity conjecture" states that if a graph G has an $s - t$ connectivity pair, then it has $a + b$ edge-disjoint (s, t) paths of which a are internally vertex-disjoint. Hikoe Enomoto and Atsushi Kaneko[19] proved that, provided b is sufficiently large relative to a, a slightly stronger version of the conjecture holds.

Theorem 5.8 (Enomoto and Kaneko's theorem). *If, in a graph G, (a, b) is an $s - t$ connectivity pair of positive integers with $b \geq a^2 + 1$, then G contains $a + b$ edge-disjoint paths from s to t, $a + 1$ of which are internally disjoint.*

Another related result appeared in 1935 and involved common representatives of left and right cosets in group theory. Due to Philip Hall of Cambridge University, the result in the language of set theory is known as *Hall's theorem.*[20] As a consequence of a later formulation by Paul Halmos and Herbert Vaughan in 1950, Hall's theorem has often been called the "marriage theorem". In graph theory terms it states that if we have a bipartite graph (possibly representing sets of boys and girls, or applicants and jobs), with an applicant-vertex joined to a job-vertex whenever the applicant is suitable for the job, then we can fill all of the jobs with suitable applicants if and only if, for each possible value of k, every k jobs have collectively at least k suitable applicants. Again, in the sense described above, this is a good theorem.

An extension of this problem was studied by David Gale and Lloyd Shapley in 1962.[21] They considered a situation with equal numbers of applicants and jobs, where each applicant has a ranking of the jobs and each job has a ranking of the applicants. They proved that it is then always possible to find a stable assignment in which each applicant is assigned a job, and where a matching of any applicant to any job can never be better for both than the matching provided by the assignment. This variation has been applied to many real-world situations—for example, allocating medical graduates to their first intern appointments. In 2012, Shapley shared a Nobel Prize in Economics for his contributions in this area.

5.2.3 1-factors. A set of independent edges in a graph is called a *matching* and (as mentioned earlier) a set of independent edges that meet all vertices (that is, a 1-factor) is a *perfect matching*. A condition for the existence of such a matching was first presented in 1947 by W. T. Tutte, who gave an algebraic proof that involved the theory of determinants.[22]

Theorem 5.9 (Tutte's theorem). *Every graph G has either a perfect matching or a set S of vertices for which $G - S$ has more than $|S|$ connected components with an odd number of vertices (but not both).*

Another criterion, which had already appeared in Petersen's paper, is that a set F of independent edges in a graph is a maximum such set if and only if there is no "alternating path"—that is, a path between two vertices, neither of which is at an end of an edge in F, where the edges alternate between those not in F and those in F. This result was also stated and proved by Claude Berge.[23]

In contrast to the Petersen–Berge theorem, Tutte's characterization shows the existence of reasons that are easily checkable for the existence and for the nonexistence of a perfect matching. Edmonds has called theorems of this type "existentially polytime", or simply "good", as explained earlier. In 1965, he conjectured that for such a theorem there is a polynomial-time algorithm for finding what must exist, and in a classic paper he boosted the theory of polynomial algorithms by presenting an efficient algorithm that, for any input graph, produces either a perfect matching or a set S, as described in Tutte's theorem.[24]

In 1950, Tibor Gallai published a paper containing a beautiful proof of Tutte's 1-factor theorem, based on a general theory of alternating paths.[25] His method splits the vertices of any graph into four classes, and the same decomposition arises from Edmonds's algorithm of 1965; the result is now called the *Gallai–Edmonds decomposition*. An interesting footnote in Gallai's paper mentions that he had also found a criterion for the existence of a 2-factor, but unfortunately he never published it. With his deep insight and thorough style, it would have been interesting to see his method.

In 1952, Tutte published his *f-factor theorem* in which, for each vertex v of a graph, an integer $f(v)$ is given with $0 \leq f(v) \leq \deg v$, and one asks for a spanning subgraph in which each vertex v has degree $f(v)$.[26] A special case provides a criterion for when a given graph has a k-factor, for any given integer k. Then, in 1954, he produced a beautiful argument that shows how the general result can actually be deduced from the 1-factor theorem.[27] This implies that Edmonds's efficient algorithm for the 1-factor problem also provides an efficient algorithm for the much more general f-factor problem.

In fact, the general k-factor problem had already been solved in the years immediately after Tutte's 1-factor paper appeared. In 1949, Hans-Boris Belck, age 19, handed in a Master's thesis at the University of Frankfurt solving the k-factor problem; it contained so many novel ideas that, on presenting a new thesis in the following year, he was awarded a Ph.D. degree. His method was to consider structural properties of maximal graphs with no k-factor, and thereby to deduce necessary and sufficient conditions for the nonexistence of a k-factor—this included the first purely graph-theoretic proof of Tutte's theorem. A similar approach was later used by László Lovász to give a very short and elegant demonstration of the 1-factor theorem.[28]

Belck published his general k-factor theorem in 1950,[29] but he then disappeared from the scene, emigrating to the United States but failing to gain mathematical employment there. Instead, he got a second Ph.D. degree (in physics) from Rensselaer Polytechnic Institute in New York and thereafter moved to Brazil where he founded a successful electronics company. He died in 2007, with his youthful graph-theoretic achievements remaining largely unrecognized by the mathematical community. One wonders whether Gallai chose not to publish his 2-factor theorem because he discovered that Belck had already done so.

In 1998, Denis Hanson, Christina Loten, and Bjarne Toft used the 2-factor theorem to extend the theorem of Petersen on cubic graphs to the general odd-regular case.[30] Their result indicates that one should regard part (b) of Petersen's theorem on cubic graphs as a 2-factor theorem.

Theorem 5.10 (Hanson, Loten, and Toft's theorem). *Every $(2k + 1)$-regular graph with at most $2k$ bridges can be split into a 2-factor and a $(2k - 1)$-factor.*

The $(2k + 1)$-regular graphs with exactly $2k + 1$ bridges but no proper factors have also been determined, thereby generalizing the Sylvester graph in Figure 5.4.[31]

Many other results resemble those above. For example, we can drop the condition that the factors should be spanning subgraphs and ask when a given graph G can be written as an *H-factorization*—that is, as the edge-disjoint union of copies of another graph H; clearly, the number of edges of G must be a multiple of the number of edges of H. Furthermore, if G is the complete graph on n vertices and the vertex-degrees in H are $d_1, d_2, \ldots, d_r$, then the equation $d_1 x_1 + d_2 x_2 + \ldots + d_r x_r = n - 1$ must have a solution in nonnegative integers $x_1, x_2, \ldots, x_r$. When n is large enough, these necessary conditions are also sufficient—a surprising result that was proved by Richard Wilson in 1975.[32] His elegant proof involved finite fields, finite vector spaces, and designs.

Wilson's paper was followed by many further investigations and results. For example, in 1989 Roland Häggkvist proved that, for any given bipartite graph H, there is a number r for which the complete bipartite graph $K_{r,r}$ has an H-factorization.[33] János Barát and Carsten Thomassen conjectured that, for any tree T, every graph G with sufficiently high edge-connectivity (depending on T) whose number of edges is a multiple of the number of edges in T, is the edge-disjoint union of copies of T. This result was first obtained in special cases, such as when T is a path, but eventually the full conjecture was proved in 2017 by J. Bensmail et al.[34]

5.3 Conclusion

The factorization problem, and the more general problem of decomposing graphs into edge-disjoint unions of particular graphs, lie at the heart of graph theory. One of the earliest purely graph-theoretic problems was that of locating a spanning 2-regular subgraph. A special type of 2-factor is a Hamiltonian cycle, but whereas the general 2-factor problem has satisfactory solutions in the form of a good theorem and a good algorithm, the Hamiltonian problem has neither.

As we have mentioned, Tutte showed in 1954 that the 2-factor problem (and more generally the f-factor problem) can be reduced to the 1-factor problem. Later, in 1965, the 1-factor problem initiated the theory of efficient algorithms through the work of Edmonds. We may wonder why a solution to the 1-factor problem came so late in the development of graph theory. It might indeed have been within the reach of Sylvester and Petersen around 1890, but they were focused on 2-factors and considered only regular graphs, rather than graphs in general. The area of splitting graphs and digraphs into special subgraphs is still an active part of graph theory today.

Further Reading

Most standard textbooks in graph theory have chapters devoted to the topics of this chapter. One such comprehensive text is

> Béla Bollobás, *Extremal Graph Theory*, Academic Press (1978); reprinted
> Dover (2004).

For a book (in German) devoted to cycles and their connections with other graph-theoretic concepts, see

> H. Walther and H.-J. Voss, *Über Kreise in Graphen*, VEB Deutscher Ver-
> lag der Wissenschaften (1974).

R. J. Wilson, A brief history of hamiltonian graphs, *Ann. Discrete Math.* 41 (1989), 489–496.

A standard monograph on matching and factorization and their history is
L. Lovász and M. D. Plummer, *Matching Theory, Ann. Discrete Math.* 29 (1989); revised AMS Chelsea Publishing (2009).

Graph factorizations are also treated in depth in
J. Akiyama and M. Kano, *Factors and Factorizations of Graphs*, Lecture Notes in Mathematics 2031, Springer (2011).

A source for the early developments of the topics of this chapter is
N. L. Biggs, E. K. Lloyd, and R. J. Wilson, *Graph Theory 1736–1936*, Clarendon Press (Oxford) (1976, 1998).

For the history of matching and factorization, see also Chapters 5 and 6 of
Robin Wilson, John J. Watkins, and David J. Parks, *Graph Theory in America*, Princeton Univ. Press (2023).

Notes and References

[1] Tait's conjecture and Kirkman's remark appear in P. G. Tait, On the colouring of maps, *Proc. Royal Soc. Edinburgh* 10 (1878–80), 501–503 and T. P. Kirkman, Question 6610 (solution by the proposer), *Math. Quest. Solut. Educ. Times* 35 (1881), 112–116. Tutte's paper, which he wrote while still involved with his wartime codebreaking, was W. T. Tutte, On Hamiltonian circuits, *J. London Math. Soc.* 21 (1946), 98–101.

[2] H. Sachs, Ein von Kozyrev und Grinberg angegebener nicht-Hamiltonscher kubischer planarer Graph, *Beiträge zur Graphentheorie, Koll. Manebach 1967*, Leipzig (1968), 127–130. Grinberg published his criterion and examples in Russian in the *Latvian Math. Yearbook* 4, Riga (1968), 51–58: for any planar graph with a Hamiltonian cycle H, $\Sigma(k-2)(f(k) - g(k)) = 0$, where $f(k)$ and $g(k)$ are the numbers of k-sided faces inside and outside H. If this can never hold, then the graph is not Hamiltonian.

[3] A. Kowalewski, W. R. Hamilton's Dodekaederaufgabe als Buntordnungsproblem, and Topologische Deutung von Buntordnungsproblemen, *Sitzungsberichte Akad. Wiss. Wien* 126 (1917), 67–90, 963–1007. For details of the books on A. Sainte-Laguë and König, see Chapter 1.

[4] L. Rédei, Ein kombinatorisher Satz, *Acta Sci. Math. (Szeged)* 7 (1934), 39–43. In notes from Aarhus University in the 1970s, G. A. Dirac proved a related stronger version of Rédei's theorem:

From a complete nonoriented graph, obtain a mixed graph by orienting some (but not all) of the edges. Then the number of Hamiltonian paths containing at least one nonoriented edge and at least one oriented edge is even.

This theorem implies that reversing the orientation of just one edge of a tournament leaves the parity of the number of Hamiltonian paths unchanged; because a transitive tournament has exactly one Hamiltonian path, Rédei's theorem follows.

[5] P. Camion, Chemins et circuits hamiltoniens des graphes complets, *C. R. Acad. Sci. Paris* 249 (1959), 2151–2152. See also F. Harary and L. Moser, The theory of round robin tournaments, *Amer. Math. Monthly* 73 (1966), 231–246; J. Moon, *Topics on Tournaments*, Holt, Rinehart and Winston (1968).

[6] G. A. Dirac, Some theorems on abstract graphs, *Proc. London Math. Soc.* (3) 2 (1952), 69–81.

[7] O. Ore, Note on Hamilton circuits, *Amer. Math. Monthly* 67 (1960), 55. Ore was of Norwegian descent, but emigrated to the United States to take up a professorship at Yale University. His 1962 book, *Theory of Graphs* (Amer. Math. Soc.), was the first graph theory textbook to

have its earliest version in English. However, prior to that, Claude Berge published his text-book *Théorie des graphes et ses applications* in French in 1958 (Dunod). This was followed by an English translation by Alison Doig in 1962.

[8]J. A. Bondy and V. Chvátal, A method in graph theory, *Discrete Math.* 15 (1976), 111–135.

[9]G. Sabidussi, Correspondence between Sylvester, Petersen, Hilbert and Klein on invariants and the factorization of graphs 1889–1891, *Discrete Math.* 100 (1992), 99–155. This paper is an annotated and translated version (from French and German) of the entire correspondence from Sylvester to Petersen in 1889–90, including some related correspondence with Hilbert and Klein. The letters from Petersen to Sylvester seem not to have survived.

[10]Petersen spent a week with Sylvester in Oxford before going with him to London, where he also met Arthur Cayley and Thomas Archer Hirst.

[11]For Sylvester's introduction of the word "graph" and his paper in *Amer. J. Math.*, see Chapter 1, Notes 11 and 12.

[12]J. Petersen, Die Theorie der regulären graphs, *Acta Math.* 15 (1891), 193–220. Note that Petersen used the English word "graphs" in the title of his otherwise German paper, having learned it from Sylvester.

[13]J. Petersen, [Sur le théorème de Tait], *Interméd. Math.* 5 (1898), 225–227.

[14]A. B. Kempe, A memoir on the theory of mathematical form, *Phil. Trans. Royal Soc. London* 177 (1886), 1–70.

[15]D. König, Graphok és matrixok, *Mat. Fiz. Lapok* 38 (1931), 116–119; this appears with a German summary entitled "Graphen und Matrizen".

[16]K. Menger, Zur allgemeinen Kurventheorie, *Fund. Math.* 10 (1927), 96–115. Menger was Austrian and moved to the United States in 1937, where he was a professor first at the University of Notre Dame and after World War II at the Illinois Institute of Technology.

[17]There are many variations of Menger's theorem, two elementary versions of which involve the number of paths between two nonadjacent vertices s and t of a graph G. Defining the *s-t vertex-connectivity* $\kappa(s,t)$ and the *s-t edge-connectivity* $\lambda(s,t)$ to be the minimum number of vertices or edges whose deletion leaves s and t in different components of G, we have the following consequences of Menger's theorem:
(a) If $\kappa(s,t) = k$, then G has a collection of k internally disjoint $s - t$ paths;
(b) If $\lambda(s,t) = l$, then G has a collection of l edge-disjoint $s - t$ paths.

[18]L. R. Ford, Jr. and D. R. Fulkerson, Maximal flow through a network, *Canad. J. Math.* 8 (1956), 399–404. See also their classic book on the subject, *Flows in Networks*, Princeton Univ. Press (1962).

[19]H. Enomoto and A. Kaneko, The condition of Beineke and Harary on edge-disjoint paths some of which are openly disjoint, *Tokyo J. Math.* 17 (1994), 355–357.

[20]P. Hall, On representatives of subsets, *J. London Math. Soc.* 10 (1935), 26–30 = *The Collected Works of Philip Hall*, Clarendon Press (Oxford) (1988), 165–169. See also P. R. Halmos and H. E. Vaughan, The marriage problem, *Amer. J. Math.* 72 (1950), 214–215.

[21]D. Gale and L. S. Shapley, College admissions and the stability of marriage, *Amer. Math. Monthly* 69 (1962), 9–15.

[22]W. T. Tutte, The factorization of linear graphs, *J. London Math. Soc.* 22 (1947), 107–111.

[23]C. Berge, Two theorems in graph theory, *Proc. Nat. Acad. Sci. U.S.A.* 43 (1957), 842–844.

[24]J. Edmonds, Paths, trees, and flowers, *Canad. J. Math.* 17 (1965), 449–465.

[25]T. Gallai, On factorisation of graphs, *Acta Math. Akad. Sci. Hungar.* 1 (1950), 133–153.

[26]W. T. Tutte, The factors of graphs, *Canad. J. Math.* 4 (1952), 314–328.

[27]W. T. Tutte, A short proof of the factor theorem for finite graphs, *Canad. J. Math.* 6 (1954), 347–352. Tutte's proof also appears in the above-mentioned book by Bollobás.

[28]L. Lovász, Three short proofs in graph theory, *J. Combin. Theory* (B) 21 (1975), 269–271. The three results proved are Brooks's theorem and the theorems of Tutte and König.

[29]H.-B. Belck, Reguläre Faktoren von Graphen, *J. Reine Angew. Math.* 188 (1950), 228–252. Belck's advisors at the University of Frankfurt in 1948–50 were Wolfgang Franz and Ruth Moufang.

[30] D. Hanson, C. O. M. Loten, and B. Toft, On interval colourings of bi-regular bipartite graphs, *Ars Combin.* 50 (1998), 23–32.

[31] A. V. Kostochka, A. Raspaud, B. Toft, D. B. West, and D. Zirlin, Cut-edges and regular factors in regular graphs of odd degree, *Graphs Combin.* 37 (2021), 199–207.

[32] R. M. Wilson, Decompositions of complete graphs into subgraphs isomorphic to a given graph, *Proc. 5th British Combin. Conf.* (1975), 647–659.

[33] R. Häggkvist, Decompositions of complete bipartite graphs, *Surveys in Combinatorics* 1989, LMS Lecture Notes 141, Cambridge Univ. Press (1989), 115–147.

[34] J. Bensmail, A. Harutyunyan, T-N. Le, M. Merker, and S. Thomass'e, A proof of the Barát-Thomassen conjecture, *J. Combin. Theory (B)* 124 (2017), 39–55.

6

Minors, Perfect Graphs, and Extremal Graph Theory

In 1840, the German mathematician August Möbius asked his students whether five countries in a map can share common borders in pairs, and the impossibility of finding such a map has led many people over the years into believing that this proves the four-color theorem. It does not, but in 1942 Hugo Hadwiger in Switzerland proposed a general conjecture which would imply that one can delete borders from any planar map requiring five colors to obtain five new countries with common borders in pairs. Not only does this latter statement imply the four-color theorem, but it is actually equivalent to it, as proved by Klaus Wagner. Hadwiger's conjecture in general is now one of the most celebrated unsolved problems in graph theory.

Continuing themes in the development of graph theory have been the relationships between properties of a graph and various of its subgraphs and minors. A subgraph H of a graph G is an *induced subgraph* if every edge of G with both ends in H is itself an edge of H. We also recall from Chapter 3 that a *minor* of G is a graph obtained from G by the operations of deleting vertices and deleting and contracting edges. In Section 6.1 we state Hadwiger's conjecture in terms of minors.

In Section 6.2, in contrast to investigating upper bounds for the chromatic number $\chi(G)$ of a graph G, we consider lower bounds. The simplest of these is the *clique number* $\omega(G)$, defined as the number of vertices in a largest complete subgraph of G. Around 1960, Claude Berge asked for a characterization of "perfect graphs", where $\omega = \chi$ for the graph and all of its induced subgraphs. Berge's work led to the so-called *strong perfect graph conjecture*, finally resolved in 2002.

How $\omega(G)$ is related to the number of edges in a graph was first investigated by Paul Turán in 1941, setting the scene for *extremal graph theory*. The breakthrough came after World War II, when Hungarian mathematicians—in particular, Paul Erdős— developed it as a subject in its own right, with Turán's theorem as its important starting point.

The final topic of this chapter is *Ramsey theory*. One special case of this is that, for any positive integers k and l and every sufficiently large number n (depending on k

and l), every graph G on n vertices either contains K_k as a subgraph or its complement (with all of the edges not in G) contains K_l. Ramsey theory applies to more general set structures than graphs, but even the graph case gives rise to interesting problems and results.

6.1 Graph minors

Hadwiger's ideas in the early 1940s were inspired by a paper from 1937 by Klaus Wagner,[1] who in turn had been inspired by Kuratowski's characterization in 1930 of planar graphs in terms of forbidden subgraphs (see Chapter 3; here we use the terminology that a graph is a subdivision of another graph if it can be obtained by repeated subdivisions of edges).

Theorem 6.1 (Kuratowski's theorem). *A graph is planar if and only if it does not contain a subdivision of either the complete bipartite graph $K_{3,3}$ or the complete graph K_5.*

Meanwhile, as we saw in Chapter 3, Karl Menger had proved a more restricted version,[2] that every nonplanar cubic graph has a subgraph that is a subdivision of $K_{3,3}$. Wagner characterized the class of graphs that contain no subdivision of $K_{3,3}$, proving that they are all 5-colorable.[3] He next tried to follow this with a corresponding investigation for K_5; however, this latter class is complicated and still not fully understood.

In 1937, instead of considering graphs containing no subdivision of K_5, he was able to characterize those with no K_5 as a minor. Wagner's characterization implies that if all planar graphs are 4-colorable, then so are all graphs in the larger class of graphs with no K_5 as a minor. This shows that the four-color theorem is equivalent to the statement that any graph with no K_5 as a minor is 4-colorable, thereby providing a new and completely combinatorial formulation of the four-color theorem.

On December 15, 1942, at a colloquium at the University of Zürich, Hadwiger[4] conjectured that any graph with no complete graph K_k as a minor is $(k-1)$-colorable. In a letter to Wagner, dated January 14, 1961, he wrote:[5]

> It does not seem quite right to me that the conjecture now is named after me. In my Zürich lecture I just took your starting point, that contained the decisive idea, and carried it over to general chromatic numbers.

Surprisingly, the two men never met in person, even though they shared many common mathematical interests and lived relatively close to each other, Wagner in Köln and Hadwiger in Zürich.

Perhaps a more striking formulation of Hadwiger's conjecture is the following, where a *topological class* is one that is closed under the taking of minors.

Conjecture 6.1 (Hadwiger's conjecture). *Within any topological class of graphs, the largest chromatic number equals the number of vertices in a largest complete graph in the class.*

Planar graphs form a topological class, and so Hadwiger's conjecture suggests that the maximum possible chromatic number of a planar graph is 4, simply because K_4 (but not K_5) belongs to it. The truth of this form of Hadwiger's conjecture for the topological class of graphs that can be embedded on a given surface (other than the plane or sphere) follows from the map color theorem of Gerhard Ringel and J. W. T. Youngs (see Chapter 3).

Hadwiger's conjecture, in the form that any graph with no complete graph K_k as a minor is $(k-1)$-colorable, was proved for $k \leq 4$ by Hadwiger in his paper of 1943, and independently in 1951 by Gabriel Dirac.[6] For $k = 5$, it is equivalent to the four-color theorem, as described above. For $k = 6$, it is also equivalent to the four-color theorem, as was shown in 1993 in a long and complicated proof by Neil Robertson, Paul Seymour, and Robin Thomas.[7] For $k \geq 7$, the conjecture remains open.

There has also been interest in a weaker version of Hadwiger's conjecture, which suggests that any graph with no complete graph K_k as a minor is ck-colorable, for some constant c; for this and other related conjectures and results, there are two excellent surveys of Hadwiger's conjecture by Ken-ichi Kawarabayashi and Paul Seymour.[8]

An even weaker version than this, with ck replaced by an exponential function of k, was proved by Wagner[9] in 1964—and with ck replaced by $ck\sqrt{\log k}$ by Alexander Kostochka and Andrew Thomasson[10] in 1984—so a sufficiently large chromatic number that depends only on k guarantees the existence of K_k as a minor. The result of Kostochka and Thomasson deals with the number of edges in graphs on n vertices and implies the existence of a complete minor of a given size when that number is sufficiently large, with the above corollary for the chromatic number. Earlier, such exact extremal results (for small complete minors) had been obtained by Gabriel Dirac, Wolfgang Mader, and Ivan Tafteberg Jakobsen.[11]

A related conjecture of Hadwiger's was proposed by the Hungarian mathematician György Hajós, who around 1940 had solved an important geometrical problem of Hermann Minkowski.[12] Encouraged by Dénes König, he turned his interest towards the four-color problem, suggesting not only that a k-chromatic graph has K_k as a minor, but that it also has a subdivision of K_k (that is, a graph obtained from K_k by repeated subdivision of edges).

Conjecture 6.2 (Hajós's conjecture). *Every graph with no K_k as a subdivision is $(k-1)$-colorable.*

Hajós never formulated this conjecture in writing, and it is possible that he considered it only for small values of k—that is, when $k = 5$ and 6. After World War II, Hajós's idea was transferred from Budapest to London through Peter Ungar, who inspired Dirac's work in graph coloring. However, Hajós's conjecture is false for $k \geq 7$, and counterexamples were provided by Paul A. Catlin[13] in 1977 (see Figure 6.1, where a pair of parallel lines means that the corresponding pairs and triples of vertices are joined by all possible edges). However, it remains unproved for $k = 5$ and $k = 6$—in particular, does every 5-chromatic graph contain a subdivision of K_5?

Soon after Catlin discovered his examples, Paul Erdős and Siemion Fajtlowicz[14] used a probabilistic argument to prove that Hajós's conjecture fails in a very strong sense: almost all graphs are counterexamples! It was probably the apparent similarity of Hajós's conjecture to that of Hadwiger that prevented mathematicians from recognizing the fundamental difference between the two.

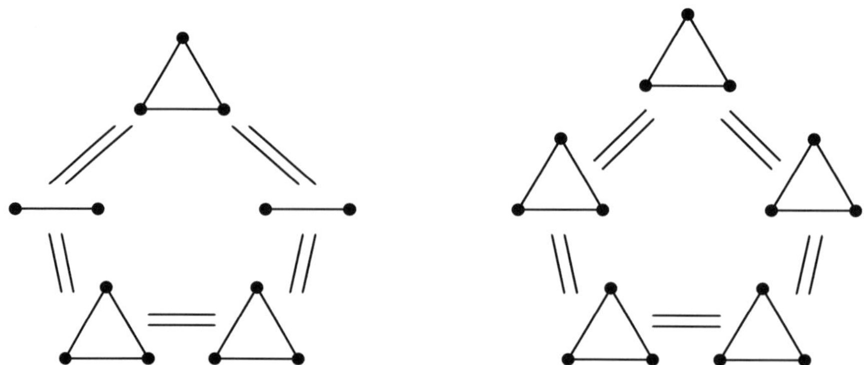

Figure 6.1. Catlin's counter-examples to Hajós's conjecture for $k = 7$ and $k = 8$

6.2 Perfect graphs

Around 1960, the French mathematician Claude Berge[15] began to look for graphs G whose chromatic number $\chi(G)$ is equal to its clique number $\omega(G)$, and where this property holds also for all of its induced subgraphs; such graphs were soon called *perfect graphs*. A conjecture about their structure, the *strong perfect graph conjecture*, was formulated, and first appeared in print in 1963 in a booklet from the Indian Statistical Institute,[16] but it may already have been mentioned by Berge at a graph theory conference in Halle in East Germany in 1960; this was only the second conference on graph theory; the first was in Dobogókő in Hungary in 1959, also with Berge participating.[17]

The odd cycles C_{2k+1} with $k \geq 2$ are not perfect since $\chi(C_{2k+1}) = 3$ and $\omega(C_{2k+1}) = 2$, but all of their proper subgraphs are. Their complements are also not perfect, because $\chi = k + 1$ and $\omega = k$, but all of their proper induced subgraphs are. It follows that odd cycles and their complements are minimal nonperfect graphs (see Figure 6.2).

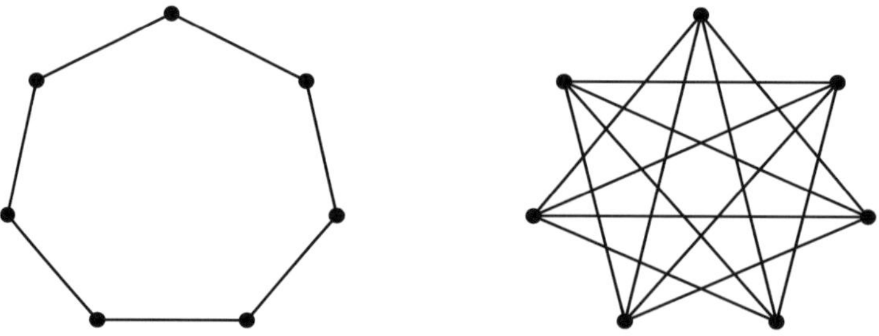

Figure 6.2. Two minimal nonperfect graphs: C_7 and its complement

The strong perfect graph conjecture suggests that these are the only such graphs, a statement that we give here in a slightly different formulation.

Conjecture 6.3 (The strong perfect graph conjecture). *A graph is perfect if and only if it contains no induced subgraph that is either an odd cycle of length 5 or greater or the complement of such a cycle.*

The truth of this conjecture implies the following result, proved in 1972 by László Lovász.[18]

Theorem 6.2 (The weak perfect graph theorem). *The complement of every perfect graph is perfect.*

His proof, derived from a result by Delbert R. Fulkerson,[19] is based on the duality theorem of linear programming. A graph is called *pluperfect* if it is perfect and if every graph obtained from it by replacing vertices by complete graphs and completely joining such neighboring complete graphs is also perfect. Fulkerson proved that the complements of pluperfect graphs are also pluperfect. Lovász then proved that the class of pluperfect graphs is identical with the class of perfect graphs, and in addition he obtained a characterization of perfect graphs that directly implies the weak perfect graph conjecture. The characterization of Lovász says that a graph is perfect if and only if every induced subgraph H satisfies $\alpha(H)\omega(H) \geq |V(H)|$, where ω is the maximum number of vertices in a complete subgraph (the *clique number*) and α is the maximum number of vertices in a subgraph with no edges (the *independence number*).

A proof of the strong perfect graph conjecture was announced in 2002 and published in 2006. This *tour de force*, a major achievement in mathematics, was carried through by Maria Chudnovsky, Neil Robertson, Paul Seymour, and Robin Thomas, and Seymour has described their struggle.[20] Their efforts yielded also a simpler and more systematic proof of the four-color theorem, as well as a solution to the graph minor problem (see Chapter 3).

6.3 Extremal graph theory

In 1907, the Dutch mathematician W. Mantel raised and answered the following question.

Question 6.3 (Mantel's question). Given a number of points, no four of which lie in the same plane, how many lines can be drawn between these points without forming any triangles?

In graphical terms, this problem sought the maximum number of edges in a simple graph with n vertices and no triangles (3-cycles). Solutions were also given by W. A. Wythoff and others:[21] the maximum number is $\frac{1}{4}n^2$ when n is even, and $\frac{1}{4}(n^2 - 1)$ when n is odd, and these bounds are achieved when the graph is a complete bipartite graph with an equal or nearly equal number of vertices in the two parts (see Figure 6.3). Mantel's problem and its solution were ignored for over thirty years.

In 1940, as mentioned in Chapter 3, Paul Turán was sent from Budapest to a labor camp in Transylvania, where he worked on building railroads. Although laboring outside for the whole day without pen and paper, he was still able to think about mathematics, and shortly before his death in 1976, in a welcoming note for the newly founded *Journal of Graph Theory*, he recalled:[22]

> Then the formal extremal problem occurred to me and I immediately felt that
> here was the problem appropriate for the circumstances. I cannot properly
> describe my feelings during the next few days. The pleasure of dealing with

 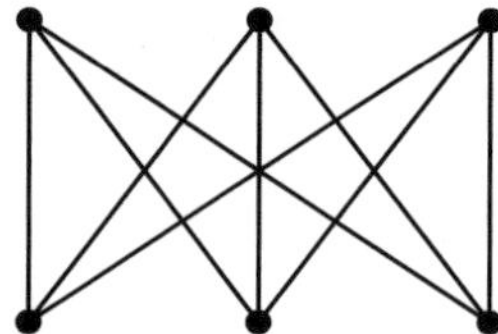

Figure 6.3. The graphs on 5 and 6 vertices with no triangles and as many edges as possible

a quite unusual type of problem, the beauty of it, the gradual nearing of the solution, and finally the complete solution, made these days really ecstatic. The feeling of some intellectual freedom and being, to a certain extent, spiritually free of oppression only added to the ecstasy.

The problem that Turán had solved was to find, for each k, the smallest number $f(n, k)$ for which any graph with n vertices and more than $f(n, k)$ edges contains K_k as a subgraph.[23] He also characterized the "extremal graphs", those graphs with n vertices and exactly $f(n, k)$ edges that contain no K_k. As noted above, the case of $k = 3$ had already been answered in 1907.

In 1946, Paul Erdős and A. H. Stone[24] proved that if a graph on n vertices has just εn^2 more edges than those given by Turán's bound for the existence of a K_k-subgraph (where ε is small and positive), then the graph contains a complete k-partite graph in which each of the k classes contains at least some given number t of vertices (for n sufficiently large, depending on ε and t). In 1966, Erdős and his compatriot Miklós Simonovits[25] pointed out that this result implies that the asymptotic behavior of the best possible function $f(n, G)$, for which any graph on n vertices and more than $f(n, G)$ edges contains G as a subgraph, depends only on the chromatic number $\chi(G)$ of G.

Theorem 6.4 (The Erdős–Stone–Simonovits theorem). *For the function* $f(n, G)$,

$$\lim_{n \to \infty} \frac{f(n, G)}{n^2} = \frac{\chi(G) - 2}{2\chi(G) - 2}.$$

So, for $\chi(G) \geq 3$, the asymptotic behavior of $f(n, G)$ is known. For 2-chromatic graphs, this theorem gives 0 as the limit, and the order of magnitude of $f(n, G)$ does not follow. A surprising consequence is that the number of edges that guarantees an even cycle with $2k$ vertices (where $\chi = 2$) is much lower than the number of edges that guarantee an odd cycle with $2k + 1$ vertices (where $\chi = 3$).

For many years extremal graph theory was dominated by the Hungarian school— it may even be said that the topic was created by Paul Erdős and his collaborators. In its classic form it is concerned with the number of edges, the vertex-degrees, and the connectivity parameters, that guarantee given properties of graphs and, whenever possible, with the determination of the extremal cases.

6.4 Ramsey theory

It is a simple matter to prove that, if we color the edges of the complete graph K_6 red and blue in any way whatever, then there must always be a monochromatic triangle: this means that for any graph G on six vertices, either $\omega(G) \geq 3$ or $\alpha(G) \geq 3$. The same statement does not hold for K_5, as we can see by coloring one 5-cycle red and the complementary 5-cycle blue. This result can be expressed informally in this way:

In any gathering of at least six people there are always three mutual friends or three mutual nonfriends, but not in any gathering of only five people.

This idea can readily be extended. To guarantee four mutual friends or four mutual nonfriends we need at least 18 people. But how many people do we need in order to guarantee five mutual friends or five mutual nonfriends? Surprisingly, the answer to this is unknown, and there are now (in 2024) four possible values, from 43 to 46.

Ramsey theory deals with the following more general type of result:

If a certain structure is divided into a certain number of classes, then at least one of the classes has a specified property.

Early in the 20th century, results of this type appeared in various forms. An example from 1916 is the theorem of Issai Schur,[26] that if the set $\{1, 2, \ldots, n\}$ is divided into t classes for a given number t and sufficiently large n, then at least one class must contain numbers $x, y,$ and z with $z = x + y$. A related example from 1927 is due to B. L. van der Waerden[27] and concerns the existence of arithmetic progressions of given length in the set of positive integers. Ramsey theory developed from such results and now forms a significant branch of graph theory.

The subject of Ramsey theory derives its name from the English economist and philosopher Frank Plumpton Ramsey, who in 1930 published his only mathematical paper, addressing a problem in logic.[28] Ramsey formulated it in an infinite version, but in its finite form it says the following.

Theorem 6.5 (Ramsey's theorem). *If the set of all subsets of size p of a set of n elements is divided into t classes, then at least one of the classes must contain all p-subsets of k of the elements, provided that n is large enough (depending on p, t, and k).*

The case $p = 2$ tells us that if the edges of the complete graph K_n are colored with t colors, then somewhere there is a monochromatic K_k provided n is large enough. As we saw above, when $t = 2$ and $k = 3$ we need only $n = 6$ vertices to force a monochromatic triangle, and because $n = 5$ is insufficient we express this by saying that the Ramsey number $r(3, 3) = 6$. More generally, the *Ramsey number* $r(k, l)$ is the smallest number for which any graph on $r(k, l)$ vertices either contains a complete subgraph K_k or its complement contains a complete subgraph K_l. Ramsey's theorem guarantees the existence of this number $r(k, l)$; in particular, for any numbers k and l, $r(2, k) = k$ and $r(l, 2) = l$.

The above result of Schur follows from Ramsey's theorem. For, take a complete graph K_n with vertices $\{1, 2, \ldots, n\}$, where n is large compared with t. Divide the vertices into t classes numbered $\{1, 2, \ldots, t\}$, and color each edge vw with the number i if the numerical difference between v and w lies in class i. A monochromatic triangle then gives the desired result.

In 1935, Paul Erdős and George Szekeres answered a problem by Esther Klein (Esther Szekeres after 1937) on point sets in the plane.[29]

Problem 6.1 (Esther Klein's problem). Given a positive integer k and n points in the plane with no three in a line and n sufficiently large, must there always be k points that form a convex polygon?

A positive answer was obtained using Ramsey's theorem. For n given points in the plane, divide the subsets of four points into two classes according to whether or not they form a convex polygon. If n is sufficiently large (depending on k), then there are either k points for which any four form a convex polygon, or there are k points for which no four form a convex polygon. But any five points in the plane already contain a convex subset of four points, and so the second case cannot occur. Therefore, in the first case, all k points must form a convex set.

Erdős and Szekeres also obtained this recursion for Ramsey numbers:

$$r(k, l) \leq r(k - 1, l) + r(k, l - 1),$$

from which we can deduce Ramsey's theorem for the case $p = t = 2$ by iteration, starting with the earlier results $r(2, k) = k$ and $r(l, 2) = l$. From this it then follows that

$$r(k, l) \leq \binom{k + l - 2}{k - 1}.$$

For Esther Klein's problem this provides an excessively large bound for n, whereas Erdős and Szekeres conjectured that the best possible value for n is $2^{k-2} + 1$; the truth of this is still unresolved.

In 1953, Ramsey numbers were studied by Robert E. Greenwood and Andrew M. Gleason.[30] Apparently without knowing the paper by Erdős and Szekeres, they obtained the same recursion and used it to obtain a table for small parameters, observing,

> ... deep combinatorial questions are encountered in attempting evaluation of the minimum. In this paper proof or disproof of the possibility of a chromatic graph construction leans heavily upon finite field theory.

In particular, they obtained the values $r(3, 4) = 9, r(3, 5) = 14$, and $r(4, 4) = 18$. As we remarked earlier, the exact value of $r(5, 5)$ is still unknown.

In 1941, Turán became aware of Ramsey's theorem and the problem of finding $r(k, k)$. In late 1944, the deportation to concentration camps was a daily happening and a threat for the Jews in Budapest. Under these extremely difficult conditions, Turán started to investigate $r(k, k)$. He immediately noticed that $(k - 1)^2$ is a lower bound, and he tried to prove that ck^2 is an upper bound for some constant c. These attempts failed, but they did have a positive effect. In 1976, Turán explained:

> Perhaps the ugly reality was what made me believe in the strong connection of beauty and truth. But the unsuccessful fight gave me strength hence, when it was necessary, I could act properly. In one of my first letters to Erdős after the war I wrote of this conjecture to him. In his answer he proved that my conjecture was utterly false.

The argument that Erdős sent to Turán and later published[31] is simple, but surprising. Erdős considered certain random objects, proved that the probability of having a particular property was positive, and deduced that an object with that property must exist. This is a purely nonconstructive existence proof—it gives no clue of how to construct the desired object. The method has since developed into a standard tool in graph theory (see Chapter 6), and has been called "Erdős magic", or more formally, the *probabilistic method*.

Instead of working with probabilities, we can often use simple counting. For a given graph G with vertices $1, 2, \ldots, n$, each pair of vertices is either an edge of G or is not. Thus there are two possibilities for each of the $\binom{n}{2}$ pairs, and therefore the number of such labeled graphs is

$$2^{\binom{n}{2}}.$$

For a given subset of k vertices, the number of graphs with a complete graph K_k on these vertices is

$$2^{\binom{n}{2}-\binom{k}{2}},$$

and the number of graphs in which these k vertices are independent is also

$$2^{\binom{n}{2}-\binom{k}{2}}.$$

Therefore, the number of graphs either with K_k as a subgraph or with k independent vertices is at most

$$\binom{n}{k}2^{\binom{n}{2}-\binom{k}{2}+1}.$$

A simple calculation shows that if $n = 2^{k/2}$, then this number is strictly less than the total number of graphs. Hence, there is a graph on $2^{k/2}$ vertices with no complete subgraph K_k and no set of k independent vertices, and this means that $r(k, k) > 2^{k/2}$ — Erdős magic! The recursion of Erdős and Szekeres and of Greenwood and Gleason implies the upper bound 2^{2k-2} for $r(k, k)$, and so $r(k, k)$ increases exponentially with k.

We saw earlier that $r(k, 2) = k$ and $r(2, l) = l$, but even the determination of $r(3, l)$ has proved difficult. The above recursion shows that $r(3, l) \leq \binom{l+1}{2} = l(l + 1)/2$. In 1983, James B. Shearer used a method of M. Ajtai, J. Komlós, and E. Szemerédi to obtain the better upper bound $r(3, l) < cl^2/\log l$, and in 1995 Jeong Han Kim proved this to be best possible, up to the constant.[32] Sam Mattheus and Jacques Verstraete recently determined the asymptotic behavior of $r(4, l)$.[33]

6.5 Conclusion

The areas of graph theory discussed in this chapter are among those having received much attention in the 20th century. The connections between complete graphs and graph properties in general still fascinate, with Hadwiger's conjecture as perhaps the most famous unsolved problem. Graph minors first played a role in Wagner's papers in the 1930s and are now at the core of graph theory, due to the deep and extensive graph minor project of Robertson and Seymour that we touched upon here and in Chapter 3, and that we shall mention again in Chapter 8.

Further Reading

For overviews of extremal graph theory, see the following:

B. Bollobás, *Extremal Graph Theory*, Academic Press (1978); reprinted Dover (2004);

B. Bollobás, Extremal graph theory, *Handbook of Combinatorics*, Vol. II (eds. R. L. Graham, M. Grötschel, and L. Lovász), North-Holland (1995), 1231–1292;

B. Sudakov, Recent developments in extremal combinatorics, Ramsey and Turán type problems, *Proc. Internat. Congress Math. IV* (2010), 2579–2606.

For further information on perfect graphs, consult
Perfect Graphs (ed. J. L. Ramirez Alfonsin and B. A. Reed), Wiley (2001).

For further reading on Ramsey graph theory, see the following:
R. L. Graham, B. L. Rothschild, and J. H. Spencer, *Ramsey Theory* (2nd edn.), Wiley (1990);

R. Graham, Old and New Problems and Results in Ramsey Theory, *Horizons of Combinatorics* (eds. E. Győri, G. O. H. Katona, L. Lovász, and G. Sági), *Bolyai Society Math. Studies* 17, Springer (2008), 105–118.

Notes and References

[1]K. Wagner, Über eine Eigenschaft der ebenen Komplexe, *Math. Annalen* 114 (1937), 570–590.

[2]K. Menger, Über plättbare Dreiergraphen und Potenzen nichtplättbarer Graphen, *Anzeiger der Akademie der Wissenschaften in Wien* 67 (1930), 85–86.

[3]K. Wagner, Über eine Erweiterung eines Satzes von Kuratowski, *Deutsche Math.* 2 (1937), 280–285. Wagner's result is that a maximal graph with no subdivision of $K_{3,3}$ is either a maximal planar graph or K_5, or is composed of such graphs by identifying complete subgraphs, and so is 5-colorable.

[4]H. Hadwiger, Über eine Klassifikation der Streckencomplexe, *Vierteljschr. Naturforsch. Ges. Zürich* 88 (1943), 133–142.

[5]This letter was shown to Bjarne Toft by Wagner and is quoted more fully in B. Toft, A survey of Hadwiger's conjecture, *Surveys in Graph Theory* (eds. G. Chartrand and M. Jacobson), *Congr. Numer.* 115 (1996), 249–283.

[6]G. A. Dirac, A property of 4-chromatic graphs and some remarks on critical graphs, *J. London Math. Soc.* 27 (1952), 85–92. In this paper, Dirac proved Hajós's conjecture for $k = 4$, as he had also done in his Ph.D. thesis at the University of London in 1951. At this stage he was unaware of Hadwiger's paper, but he had learned about Hajós's conjecture from Peter Ungar.

[7]N. Robertson, P. D. Seymour, and R. Thomas, Hadwiger's conjecture for K_6-free graphs, *Combinatorica* 13 (1993), 279–361.

[8]K. Kawarabayashi, Hadwiger's conjecture, *Topics in Chromatic Graph Theory* (eds. L. W. Beineke and R. J. Wilson), Cambridge Univ. Press (2015), 73–93 and P. Seymour, Hadwiger's conjecture, *Open Problems in Mathematics* (eds. J. Nash and M. Rassias), Springer (2016), 417–437. See also Toft's earlier and more historical survey cited in Note 6.1.

[9]K. Wagner, Beweis eine Abschwächung der Hadwiger-Vermutung, *Math. Annalen* 114 (1964), 139–141. A similar result for Hajós's conjecture appears in G. A. Dirac, Chromatic number and topological complete subgraphs, *Canad. Math. Bull.* 8 (1965), 711–715.

[10]A. Kostochka, Lower bound of the Hadwiger number of graphs by their average degree, *Combinatorica* 4 (1984), 307–316;
A. Thomason, An extremal function for contractions of graphs, *Math. Proc. Cambridge Philos. Soc.* 95 (1984), 261–265.

[11]G. A. Dirac, Homomorphism theorems for graphs,Math. Annalen 153 (1964), 69–80;
W. Mader, Homomorphiesätze für Graphen, *Math. Annalen* 178 (1968), 154–168;
I. T. Jakobsen, On certain homomorphism properties of graphs, I, II, *Math. Scand.* 31 (1972), 379–404 and 52 (1983), 229–261.

[12]Hajós proved by group-theoretic means that, if a Euclidean space of any dimension is tiled by hypercubes (whose positions are restricted to form a certain lattice), then some pair of hypercubes must meet face-to-face.

[13]P. A. Catlin, Hajós' graph-coloring conjecture: variations and counterexamples, *J. Combin. Theory (B)* 23 (1977), 247–250.

[14]P. Erdős and S. Fajtlowicz, On the conjecture of Hajós, *Combinatorica* 1 (1981), 141–143.

[15]C. Berge, Färbung von Graphen, deren sämtliche bzw. deren ungerade Kreise starr sind, *Wiss. Z. Martin Luther Univ. Math.-Natur. Reihe* 10 (1961), 114–115. This short note was originally submitted with the title (here in English translation) "Coloring of Gallai and semi-Gallai graphs," but the referee asked Gallai who modestly answered that he had not looked carefully at these classes, and so it received its more cumbersome title, suggested by Dirac.

[16]C. Berge, Perfect graphs, *Six Papers on Graph Theory*, Indian Statistical Institute, Calcutta (1963), 1–21. This seems to be the first published version of the perfect graph conjecture. For a more detailed history of the conjecture, see C. Berge, Motivations and history of some of my conjectures, *Discrete Math.* 165/166 (1997), 61–70 and B. Toft, Claude Berge – sculptor of graph theory, *Graph Theory in Paris* (from a conference in memory of Claude Berge, GT04), Birkhäuser (2007), 1–9.

[17]There are no proceedings from the Dobogókő meeting, but there is a short and interesting note about it: W. T. Tutte, A colloquium on graph theory, *Canad. Math. Bull.* 3 (1960), 102–103.

[18]L. Lovász, A characterization of perfect graphs, *J. Combin. Theory* (B) 13 (1972), 95–98; reprinted in *Classical Papers in Combinatorics* (eds. I. Gessel and G.-C. Rota), Birkhäuser (1987), 49–59; see also L. Lovász, Normal hypergraphs and the perfect graph conjecture, *Discrete Math.* 2 (1972), 253–267.

[19]D. R. Fulkerson, The perfect graph conjecture and pluperfect graph theorem, *Proc. Second Chapel Hill Conf. on Combin. Math. and Applications* (eds. R. C. Bose et al.) (1970), 171–175.

[20]M. Chudnovsky, N. Robertson, P. Seymour, and R. Thomas, The strong perfect graph theorem, *Annals of Math.* 164 (2006), 51–229;
P. Seymour, How the proof of the strong perfect graph conjecture was found, *Gaz. Math.* 109 (2006), 69–83.

[21]W. Mantel and W. A. Wythoff, [Solution to problem 28], *Wisk. Opg. Wisk. Genoot* (New Ser.) 10 (1906–10), 60–61.

[22]P. Turán, A note of welcome, *J. Graph Theory* 1 (1977), 7–9.
Founded by Frank Harary, the *Journal of Graph Theory* was the first journal devoted exclusively to the field of graph theory.

[23]P. Turán, An extremal problem in graph theory (in Hungarian), *Mat. Fiz. Lapok* 48 (1941), 436–452.
An English translation of this paper appeared in
Collected Papers of Paul Turán, Vol. 1 (ed. P. Erdős), Akadémiai Kiadó Budapest (1990), 231–240.
Informative notes about Turán's paper (by M. Simonovits) appeared in the same volume.

[24]P. Erdős and A. H. Stone, On the structure of linear graphs, *Bull. Amer. Math. Soc.* 52 (1946), 1087–1091.

[25]P. Erdős and M. Simonovits, A limit theorem in graph theory, *Stud. Sci. Math. Hungar.* 1 (1966), 51–57.

[26]I. Schur, Über die Kongruenz $x^m + y^m \equiv z^m (\bmod\ p)$, *Jber. Deutsch. Math.-Verein.* 25 (1916), 114–117.

[27]B. van der Waerden, Beweis einer Baudetschen Vermutung, *Nieuw Arch. Wisk.* 15 (1927), 212–216.

[28]F. P. Ramsey, On a problem in formal logic, *Proc. London Math. Soc.* 30 (1930), 264–286.

[29]P. Erdős and G. Szekeres, A combinatorial problem in geometry, *Compositio Math.* 2 (1935), 463–470. In the early 1930s, a group of mathematical friends, students of Dénes König (among them Paul Erdős, Tibor Gallai, Esther Klein, Martha Svéd, George Szekeres, and Paul Turán), used to meet regularly in the Budapest City Park in the area around the statue "Anonymus". The problem posed there by Esther Klein was later called the "happy end problem" by Erdős because Szekeres and Klein married in 1937. They fled from Budapest to Shanghai before World War II, and remained there during the Japanese occupation and the Chinese Civil War. In 1947 they emigrated to Australia, where they lived for the rest of their lives.

[30]R. E. Greenwood and A. M. Gleason, Combinatorial relations and chromatic graphs, *Canad. J. Math.* 7 (1955), 1–7.

[31]P. Erdős, Some remarks on the theory of graphs, *Bull. Amer. Math. Soc.* 53 (1947), 292–294.

[32]The papers mentioned in this paragraph are

- J. B. Shearer, A note on the independence number of a triangle free graph, *Discrete Math.* 46 (1983), 83–87.

- M. Ajtai, J. Komlós, and E. Szemerédi, A note on Ramsey numbers, *J. Combin. Theory* (*A*) 29 (1980), 354–360.

- J. H. Kim, The Ramsey number $r(3, t)$ has order of magnitude $t^2/\log t$, *Random Structures and Algorithms* (1995), 173–207.

[33]S. Mattheus and J. Verstraete, The asymptotics of $r(4, t)$, *Ann. of Math.* 199 (2024), 919–941. The two authors also published an interesting survey of the recent development in Ramsey theory: J. Verstraete and S. Mattheus, Off-diagonal Ramsey numbers from pseudorandom graphs, *SIAM News* 57 (2024), 1–2.

7

Graph Enumeration and Probability

Combinatorial mathematics deals with the existence and characterization of objects with a certain property, such as graphs that can be embedded in a given surface, or trees that span a given graph. In situations where their existence is known, we may ask for the number of ways in which this is realized. Several techniques have been developed to attack such enumeration problems, such as binomial identities, recurrence relations, generating functions, the inclusion-exclusion principle, matrix methods from linear algebra, and double-counting methods (where we show the equality of two expressions by counting the same set of objects in two different ways).

This chapter opens with Arthur Cayley's enumerations of rooted and unrooted trees, and with Heinz Prüfer's proof of Cayley's formula for the number of labeled trees with a given number of vertices. We continue with George Pólya's theory of counting graphs of various types, where there is a degree of symmetry; related results had been obtained a few years earlier by J. Howard Redfield, but they remained largely unknown, partly because of Redfield's complicated choice of terminology.

Rather than counting directly, Paul Erdős formulated results in probabilistic terms, developing a powerful probabilistic method, referred to in Chapter 6 as "Erdős magic". Around 1960, with his compatriot Alfréd Rényi, he subsequently created the theory of random graphs and their evolution.

As we shall see in Section 7.2, the probabilistic method and the study of random graphs have developed into a rich and fruitful theory that has had a substantial impact for graph theory as a whole—for example, a striking result of Béla Bollobás determines the chromatic number of random graphs in a very precise way.

We conclude the chapter with some general results. We first present László Lovász's local lemma and Endre Szemerédi's regularity lemma, both of which are extensively used in combinatorial theory. We conclude with a wonderful "fake" game-theoretic adaptation of the probabilistic method, due to József Beck, which presents Erdős magic in a new form. For large graphs where the error terms may be negligible, exact numerical results can sometimes be obtained.

The results of this chapter bear witness to the extremely strong influence of Hungarian mathematicians in the development of combinatorial mathematics and its connections with probability and statistics.

7.1 Enumeration

As we saw in Chapter 1, a *tree* is a connected graph without cycles. A tree with n vertices is *rooted* if one particular vertex is designated as the "root", and is *unrooted* otherwise. It is *labeled* if its vertices are numbered—for example, from 1 to n.

The first mathematical use in print of the word "tree" was due to the English mathematician Arthur Cayley in 1857, although the concept had arisen ten years earlier in Gustav Robert Kirchhoff's fundamental work on electrical networks. In the 1850s, Cayley became interested in counting rooted and unrooted trees, with applications to the enumeration of certain chemical molecules, and wrote several papers on the subject. He summarized his results in a short paper that was published in 1881.[1]

Another important result in graph enumeration is Cayley's formula for the number of labeled trees with n vertices,[2] which he presented in 1889.

Formula 7.1 (Cayley's formula). *The number of labeled trees with n vertices is n^{n-2}.*

Thus, for four vertices there are sixteen labeled trees, as shown in Figure 7.1.

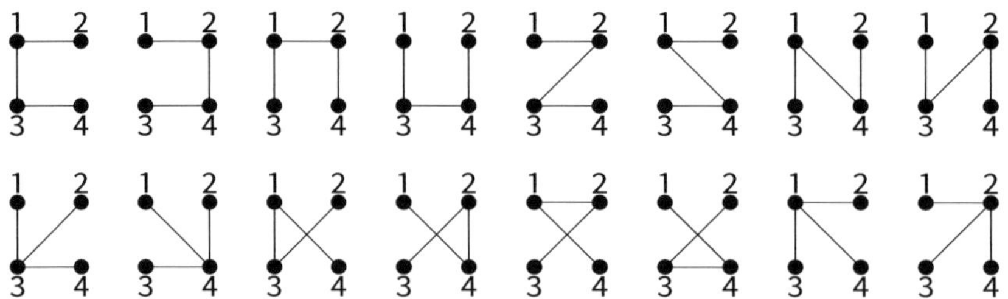

Figure 7.1. The sixteen labeled trees with four vertices

There are many different proofs of Cayley's formula, as presented by John Moon[3] and earlier calculations of determinants by the German mathematicians Karl Georg Christian von Staudt and Carl Wilhelm Borchardt were equivalent to it. However, Cayley proved his result only for $n = 6$, saying hopefully:

> . . . it will be at once seen that the proof given for this particular case is applicable for any value whatever of n.

In 1918, Heinz Prüfer[4] presented a correct proof for all values of n by establishing a one-to-one correspondence between labeled trees with n vertices and $(n-2)$-tuples $(a_1, a_2, \ldots, a_{n-2})$, where each a_i is an integer from 1 to n. Because there are n^{n-2} such $(n-2)$-tuples, the result follows immediately. Other results of a similar type may be found in a paper of S. L. Hakimi.[5]

7.1.1 From Redfield to Pólya. As we mentioned in the introduction to this chapter, a paper from 1927 by J. Howard Redfield was concerned with the enumeration of objects when a degree of symmetry is involved.[6] How many essentially different objects are there, if two objects are regarded as the same whenever we can transfer from one to the other by a symmetry, such as a rotation or a reflection? For example:

In how many essentially different ways can we color the eight corners of a cube black and white, if we are allowed to rotate the cube in 3-dimensional space?

Redfield's paper was largely ignored at the time, partly because of its inaccessible language (such as a "symmetrical aliorative dyadic relation number" for what we call a simple graph).

Such general problems were taken up in 1935–37 by the Hungarian mathematician George Pólya. Born in Hungary in 1887, he received his basic education there. From 1914 to 1940, he was a professor in Zürich before moving to the United States, where he remained at Stanford University in California for the rest of his life. He died in 1985. Among his many books, *How to Solve it*, published in 1945 and reissued many times, became a classic.

Unfamiliar with Redfield's work, Pólya published several papers on enumeration problems involving symmetry, including a lengthy seminal paper in 1937 that provided the algebraic foundation for their solution.[7] His main result, which we explain below, is as follows.

Theorem 7.1 (Pólya's theorem). *The number of nonequivalent labelings, with respect to a group G of symmetries, is obtained by substituting the generating function of the labels into the cycle index of G.*

To explain these terms, we consider an example that Pólya used in 1978, when at the age of 90 he lectured on his theorem in a course at Stanford.[8] The cycle graph C_6 has as its group of automorphisms the dihedral group $\mathbb{D}_6$, which has 12 elements: two 1-step rotations (clockwise and counter-clockwise) represented by 6-cycles, two 2-step rotations represented by two 3-cycles (see Figure 7.2), four reflections represented by three 2-cycles, three reflections represented by two 1-cycles and two 2-cycles, and the identity element represented by six 1-cycles. The *cycle index* arising from these elements is then defined to be

$$\tfrac{1}{12}(2s_6 + 2s_3^2 + 4s_2^3 + 3s_1^2 s_2^2 + s_1^6),$$

where, for example, the term $3s_1^2 s_2^2$ represents the three permutations consisting of two 1-cycles and two 2-cycles.

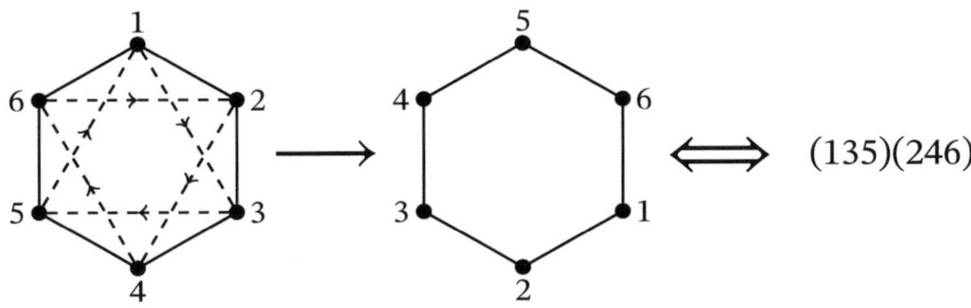

Figure 7.2. A 2-step rotation represented by two 3-cycles

Assume now that we wish to label the vertices of C_6 with one of the three labels, blue (b), green (g), and red (r). The *generating function* for these labels is $b + g + r$, with the three variables representing the colors. Substituting this into the cycle index means replacing each symbol s_i by $b^i + g^i + r^i$, giving

$$\tfrac{1}{12}\{2(b^6 + g^6 + r^6) + 2(b^3 + g^3 + r^3)^2 + 4(b^2 + g^2 + r^2)^3$$
$$+ 3(b + g + r)^2(b^2 + g^2 + r^2)^2 + (b + g + r)^6\}.$$

If we expand this expression, one of the terms is $3bgr^4$, and this means that there are exactly three nonequivalent ways to label the vertices of C_6 with one blue label, one green label, and four red labels (see Figure 7.3). Each of the other terms of the expression can be interpreted in a similar way.

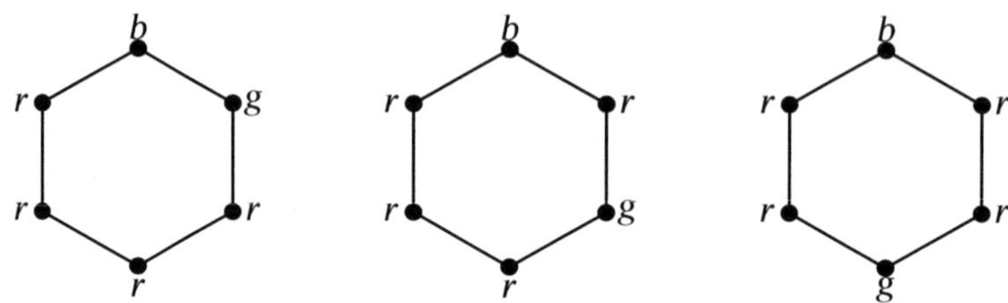

Figure 7.3. The three nonequivalent ways to label
C_6 with one blue, one green, and four red labels

Assume now that there are two different types of red labels: cheap and expensive. Replacing r by $r_c + r_e$ in the above expression we find (among other terms) the term $16bgr_c^2 r_e^2$. This means that there are exactly 16 nonequivalent ways to label C_6 with one blue label, one green label, two cheap red labels, and two expensive red labels. In Figure 7.3 the first diagram gives rise to six new ways, the second diagram also gives rise to six ways, and the third diagram gives rise to four ways—that is, to 16 ways in total.

Alternatively, taking $b + g + 2r$ as the generating function for the labels, and replacing s_i by $b^i + g^i + 2r^i$, we find (among other terms) the term $42bgr^4$. This means that there are 42 ways to label C_6 with three colors (blue, green, and red), where blue and green are used once each, and where red (used four times) comes in two varieties; the 16 ways above constitute some of these 42 ways—namely, those in which each of the two varieties is used twice. In Figure 7.3, each of the first two diagrams gives rise to 16 new ways, and the third diagram gives rise to 10 ways.

Pólya's papers had a profound influence on the progress of combinatorial enumeration, and his methods have been applied to many counting problems involving graphs and chemical molecules. In 1948, the American Richard Otter combined Pólya's approach with a more direct combinatorial one in order to count various types of trees,[9] and his methods led to asymptotic values for the coefficients of the generating functions under consideration. Later, the American graph theorist Frank Harary used Pólya's approach to count undirected graphs, directed graphs, and rooted graphs.[10]

7.2 Probability

As we saw in Chapter 6, Paul Erdős counted labeled graphs on n vertices that contain a complete k-graph K_k, and also those that contain k independent vertices. When $n = 2^{k/2}$, a simple calculation shows that these two numbers add up to less than the total number of labeled graphs on n vertices. It follows that there are graphs on $2^{k/2}$ vertices that have no complete k-graph as a subgraph and no set of k independent vertices. This argument might also be formulated in terms of probabilities: the probability that a graph on $2^{k/2}$ vertices has neither k mutually adjacent vertices nor k independent vertices is strictly positive. Such a graph must therefore exist (Erdős magic!).

This nonconstructive way of proving existence has turned out to be extremely powerful for showing the existence of graphs with certain properties, without introducing structure. In fact, constructive tools for obtaining graphs with a given property force the introduction of structure, and they thereby restrict the graphs under consideration. Once learned, the probabilistic method is straightforward to use and often simple to understand, and sometimes yields surprising results about almost all graphs—for example, that Hajós's conjecture fails for almost all graphs, as explained in Chapter 6.

The *girth* of a graph is the length of its shortest cycle; for example, the Petersen graph has girth 5. In 1959, Erdős[11] used the probabilistic method to prove the existence of graphs with large girth, chromatic number k and yet polynomial of order k. The classical constructions of k-chromatic triangle-free graphs all produce graphs of exponential order in k, except for a geometrical construction of 1958 by Erdős,[12] which gives such graphs of order at most ck^{50} for some positive constant c.

Theorem 7.2 (Erdős's theorem). *For all $h \geq 4$ and sufficiently large k, there exist k-chromatic graphs with at most k^{2h} vertices and girth at least h.*

The restriction on k being large can be removed if the exponent $2h$ in the theorem is replaced by $4h$; see, for example, the exposition by Bollobás.[13]

7.2.1 Random graphs. In 1959, and continuing in the succeeding years, Paul Erdős and Alfréd Rényi took the arguments a great step further by pubishing papers that created a probabilistic theory of random graphs that is not primarily based on counting arguments, but on probability theory and statistics. They investigated properties of graphs with n vertices and m edges selected at random, and asked how many connected components does it have? How large are they? and so forth. For many of these questions, the answers are surprising.

The Erdős–Rényi way of looking at random graphs was to consider the graphs $G(n, m)$ with n vertices and m edges as a probability space with elements of equal probability. An alternative model, due to E. N. Gilbert, consists of all graphs $G(n, p)$ with n labeled vertices, in which the edges appear independently with probability p; thus, a particular graph with n vertices and m edges appears with probability

$$p^m (1 - p)^{\binom{n}{2} - m}.$$

In the first of their papers on this topic, Erdős and Rényi studied connectivity properties of random graphs,[14] and three of their basic results are the following. Here, an *isolated vertex* is a vertex of degree 0.

Theorem 7.3 (Erdős and Rényi's theorem). *If* $m = \lfloor (n \log n)/2 + cn \rfloor$, *where c is a positive constant, then the following results hold:*

(a) *Almost all graphs with n vertices and m edges have one large connected component, and all of the other components are isolated vertices. The mean number of isolated vertices is* e^{-2c}.

(b) *The probability that a random graph on n vertices and m edges is connected has limit* e^{-k}, *where* $k = e^{-2c}$, *as* $n \to \infty$.

(c) *For any* $\varepsilon > 0$, *almost all graphs on n vertices and at least* $(n \log n)/(2 - \varepsilon)$ *edges are connected, whereas almost all graphs on n vertices and at most*

$$(n \log n)/(2 + \varepsilon)$$

edges are disconnected.

This last statement exemplifies the sudden changes that a random graph can undergo as the number of edges increases—there is a rather sharp threshold between graphs that "almost certainly do not have a property P" and graphs that "almost certainly do have the property P".

There are several such thresholds, and these formed the content of the next paper on random graphs by Erdős and Rényi.[15] As m increases (depending on n), one of the connected components suddenly becomes larger than any other component (the "giant component"), while the other components are small trees. Then, as m increases further, these other components become isolated vertices, with the giant component swallowing up those components with edges. Finally, the isolated vertices disappear, and the graph becomes connected. This lengthy paper by Erdős and Rényi was a real milestone, being the most important publication on the subject of random graphs.

Later, Béla Bollobás made major contributions to the understanding of how random graphs evolve. One striking result is his theorem on the chromatic number of a random graph.[16]

Theorem 7.4 (Bollobás's theorem). *For a fixed number p with* $0 < p < 1$, *a random graph* $G(n, p)$ *satisfies*

$$\left(\tfrac{1}{2} - o(1)\right)\left(\log \frac{1}{1-p}\right)\left(\frac{n}{\log n}\right) \leq \chi(G(n, p)) \leq \left(\tfrac{1}{2} + o(1)\right)\left(\log \frac{1}{1-p}\right)\left(\frac{n}{\log n}\right),$$

with probability approaching 1 as $n \to \infty$, *where* $o(1)$ *represents a function of n that converges to 0 as* $n \to \infty$.

Hence, if the probability for each edge is $p = \tfrac{1}{2}$, then (for large n) such a graph on n vertices has chromatic number close to $n \log 2/2 \log n$. The falsity of Hajós's conjecture (see Chapter 6) follows directly from this, because there is insufficient room within n vertices for a subdivision of the complete graph with so many vertices—approximately half of its edges would be undivided, whereas the other half would have at least one subdividing vertex, which is too many. So Hajós's conjecture fails for almost all graphs.

Three further results relate to themes from this section.

The first, due to László Lovász in 1975, has proved to be most important and useful.[17] It was one of his many outstanding contributions to combinatorial mathematics and computer science, for which he was awarded an Abel Prize in 2021.

Lemma 7.5 (Lovász's local lemma). *Let a set of (bad) events be given, each with probability at most p, where $0 < p < 1$, and with each event independent of all but at most d other events. If $4pd \leq 1$ then, with positive probability, no bad event occurs.*

The original application of Lovász's local lemma was to a problem on the coloring of hypergraphs, giving an upper bound for the number of vertices in a 3-critical *r*-uniform hypergraph in which any two edges intersect. Here, a *hypergraph* is a generalization of a graph, in which edges may consist of any number of vertices, rather than just 2, and is *r-uniform* if every edge contains exactly *r* vertices.

Using Lovász's local lemma, the Canadians Michael Molloy and Bruce Reed obtained the following result on list-colorings of graphs (see Chapter 2).[18]

Theorem 7.6 (Molloy and Reed's theorem). *If, for each vertex of a graph G, there is a list of k available colors, and if each color is available for at most $k/8$ neighbors of any one vertex, then the graph G can be properly list-colored.*

The proof starts from a random coloring, where each vertex receives a color from its list, and where a bad event, corresponding to an edge vw and a color i, occurs when v and w both receive color i; this happens with probability $p = 1/k^2$. Moreover, such a bad event depends on only $k/8$ neighbors of each of v and w and their colors—that is, on at most $d = 2k \times k/8 = k^2/4$ other bad events. Because $4pd = 1$, it follows from Lovász's local lemma that there is a positive probability that no bad event occurs, and this proves the theorem. Stronger versions of this result have also been studied—for example, Molloy and Reed have conjectured that $k/8$ can be replaced by $k - 1$.

Our next result, the *regularity lemma* of the Hungarian mathematician Endre Szemerédi,[19] is a strong tool in combinatorial mathematics; for example, it implies the Erdős–Stone–Simonovits theorem, cited in Chapter 6. In 1975, it was also a primary tool in the proof of a celebrated theorem of Szemerédi on arbitrarily long arithmetic progressions in sequences of integers with positive upper density, one of the deepest and most difficult results in combinatorial mathematics. For his fundamental contributions to discrete mathematics and theoretical computer science, and in recognition of the profound and lasting impact of these contributions to additive number theory and ergodic theory, Szemerédi was awarded the Abel Prize in 2012.

Lemma 7.7 (Szemerédi's regularity lemma). *For each $\varepsilon > 0$ and each integer $n > 1$, there is an integer N for which the vertices of any graph with at least N vertices can be partitioned into $k + 1$ parts $\{V_0, V_1, \ldots, V_k\}$, where $n \leq k \leq N$ and the partition is ε-regular.*

Here, a partition is *ε-regular* if the set V_0 has size at most εN (that is, V_0 is a small left-over set), the other k sets have equal size and, for all but at most εk^2 of the pairs (V_i, V_j), the edges between V_i and V_j are distributed in a regular way. The number N in the regularity lemma can be viewed as an upper bound on the number of sets in the partition, which implies that, for a large graph, each part is large.[20]

Our final result is less well known: it arises from combinatorial game theory and is quite surprising. In 2000, József Beck developed a game-theoretic adaptation of the probabilistic method to determine exact results for some classes of games. His monograph[21] is likely to become a milestone in the development of combinatorial game theory. In his *Maker–Breaker game* two people, Maker and Breaker, play on the complete graph K_n, alternately selecting a new edge. Maker's goal is to form a complete k-graph, whereas Breaker's is to prevent Maker from doing so. If k is small compared with n, then Maker has a winning strategy, but from some point on as k increases, Breaker can win. This turning point is known exactly when n is very large (say, $2^{10^{10}}$) but not when, for example, n is 20, 100, or 2^{100}.

7.3 Conclusion

As illustrated in this chapter, probability theory has come to play an increasingly important role in many aspects of graph theory, as arguments often depend on probabilistic methods. It has therefore become essential in studies of graph theory to take this aspect into serious consideration.

Further Reading

For a standard text on enumeration that features Pólya's methods, see

Frank Harary and Edgar M. Palmer, *Graphical Enumeration*, Academic Press (1973).

Fifty years after the publication of Pólya's pioneering paper, Ronald Read published an English translation of it, with a survey of further developments:

G. Pólya and R. C. Read, *Combinatorial Enumeration of Groups, Graphs and Chemical Compounds*, Springer (1987).

For excellent monographs on the probabilistic method, see the following:

P. Erdős and J. Spencer, *Probabilistic Methods in Combinatorics*, Academic Press (1974).

N. Alon and J. H. Spencer, *The Probabilistic Method* (4th edn.), Wiley (2008).

The standard text on coloring and probability is

M. Molloy and B. Reed, *Graph Colouring and the Probabilistic Method*, Springer (2002).

Excellent comprehensive monographs on random graphs are

B. Bollobás, *Random Graphs* (2nd edn.), Cambridge Univ. Press (2001);

S. Janson, T. Łuczak, and A. Ruciński, *Random Graphs*, Wiley (2000).

A more leisurely exposition is

E. M. Palmer, *Graphical Evolution: An Introduction to the Theory of Random Graphs*, Wiley (1985).

There are also several useful articles:

M. Karoński, Random graphs, *Handbook of Combinatorics*, Part I, North-Holland (1995), 351–380;

J. Spencer, Probabilistic methods, *Handbook of Combinatorics*, Part II, North-Holland (1995), 1785–1817;

R. J. Kang and C. McDiarmid, Colouring random graphs, *Topics in Chromatic Graph Theory*, (eds. L. W. Beineke and R. J. Wilson), Cambridge Univ. Press (2015), 199–229.

An excellent general combinatorics book, containing several counting methods only touched on here, is

P. J. Cameron, *Combinatorics: Topics, Techniques, Algorithms*, Cambridge Univ. Press (1994).

Many papers in these subjects by George Pólya and Paul Erdős have been reprinted in

R. P. Boas (ed.), *George Pólya, Collected Papers*, Vols. I, II, MIT Press (1974);

J. Spencer (ed.), *P. Erdős, The Art of Counting: Selected Writings*, MIT Press (1973).

Finally, many standard textbooks contain fine expositions of topics from this chapter, such as those mentioned in Notes 8 and 13.

Notes and References

[1]A. Cayley, On the analytical form called trees, *Amer. J. Math.* 4 (1881), 266–268 = *Collected Papers*, Vol. XI, Cambridge Univ. Press (1897), 365–367. Cayley was one of the most prolific mathematicians of all time, and his 978 papers run to 13 volumes, published in 1889–1898 by Cambridge University Press.

[2]A. Cayley, A theorem on trees, *Quart. J. Pure Appl. Math.* 23 (1889), 376–378 = *Collected Papers*, Vol XIII, Cambridge Univ. Press (1897), 26–28.

[3]J. W. Moon, Various proofs for Cayley's formula for counting trees, *A Seminar on Graph Theory* (ed. F. Harary), Holt, Rinehart and Winston (1967), 70–78. This seminar took place at University College, London in 1962–63. The book also contains chapters by Frank Harary on Pólya's enumeration theorem and by Erdős on applications of probabilistic methods to graph theory. For further information on trees, see J. W. Moon, *Counting Labelled Trees*, Canadian Math. Congress (1970).

[4]H. Prüfer, Neuer Beweis eines Satzes über Permutationen, *Archiv der Math. Phys.* 27 (1918), 142–144.

[5]S. L. Hakimi, On realizability of a set of integers as degrees of the vertices of a linear graph, *J. Soc. Indust. Appl. Math.* 10 (1962), 496–506.

[6]J. H. Redfield, The theory of group-reduced distributions, *Amer. J. Math.* 49 (1927), 433–455; E. K. Lloyd, Redfield's contributions to enumeration, *Match No.* 46 (2002), 214–233.

[7]G. Pólya, Kombinatorische Anzahlbestimmungen für Gruppen, Graphen und chemische Verbindungen, *Acta Math.* 68 (1937), 145–254. An annotated English translation is mentioned above. Strangely, this milestone in graph theory does not appear in Pólya's *Collected Papers*, which includes publications from 1913–1973. Pólya was also a keen photographer, and a selection of his photographs of mathematicians appears in G. Pólya and G. L. Alexanderson, *The Pólya Picture Album. Encounters of a Mathematician*, Birkhäuser (1987).

[8]G. Pólya, R. E. Tarjan, and D. R. Woods, *Notes on Introductory Combinatorics*, Birkhäuser, 1983. This book arises from lectures at Stanford University in 1978, taught by Pólya and Tarjan with Woods as their teaching assistant.

[9]R. Otter, The number of trees, *Ann. of Math.* (2) 49 (1948), 583–599.

[10]F. Harary, The number of linear, directed, rooted, and connected graphs, *Trans. Amer. Math. Soc.* 78 (1955), 445–463. See also *A Seminar on Graph Theory* (Note 3).

[11]P. Erdős, Graph theory and probability, *Canad. J. Math.* 11 (1959), 34–38 = *The Art of Counting: Selected Writings*, MIT Press (1973), 406–410.

[12]H. Sachs, Finite graphs, *Recent Progress in Combinatorics* (ed. W. T. Tutte), Academic Press (1969), 175–184. This paper is a survey of methods for obtaining k-chromatic graphs without short cycles. Erdős's graphs consist of points with integer coefficients in d-dimensional space inside a ball around the origin, with edges joining points that are at least a certain distance from each other.

[13]B. Bollobás, *Extremal Graph Theory*, Academic Press (1978); reprinted Dover (2004).

[14] P. Erdős and A. Rényi, On random graphs I, *Publ. Math. Debrecen* 6 (1959), 290–297 = *The Art of Counting* 561–568 (see Note 11).

[15] P. Erdős and A. Rényi, On the evolution of random graphs, *Magyar Tud. Akad. Mat. Kut. Int. Közl.* 5 (1960), 17–61; see *The Art of Counting*, 569–573.

[16] B. Bollobás, The chromatic number of random graphs, *Combinatorica* 8 (1988), 49–55.

[17] P. Erdős and L. Lovász, Problems and results on 3-chromatic hypergraphs and some related results, *Infinite and Finite Sets* (eds. A. Hajnal et al.), *Colloq. Math. Soc. J. Bolyai* 11 (1975), 609–627.
This paper was the first printed appearance of Lovász's local lemma, one of his earliest contributions to combinatorics.

[18] M. Molloy and B. Reed, *Graph Colouring and the Probabilistic Method*, Springer (2002) (listed above).

[19] E. Szemerédi, Regular partitions of graphs, *Éditions du CNRS* (1978), 399–401.

[20] Expositions of the regularity lemma and the Erdős–Stone theorem can be found in R. Diestel, *Graph Theory*, Springer (1997); and J. A. Bondy and U. S. R. Murty, *Graph Theory*, Springer (2008).

[21] József Beck, *Combinatorial Games*, Cambridge Univ. Press (2008); paperback edn. (2011).

8

Graph Algorithms and Complexity

An *algorithm* is a finite step-by-step procedure for solving mathematical problems. We can think of it as a recipe, where we start with some input, apply the steps of the recipe, and produce the required output—or as a black box, where we input some data, turn the handle, and produce an output that solves our problem. Algorithms can be traced back to around the year 1800 BC when the ancient Mesopotamians gave step-by-step procedures for solving particular instances of what we now call quadratic equations.

The word "algorithm" derives from the 9th-century Persian mathematician al-Khwārizmī, who wrote an influential arithmetic book on the Hindu–Arabic numerals. When this book was translated into Latin in the 12th century, the opening words became "Dixit algorismus" and the word *algorism* came to be used to mean "arithmetic", referring primarily to the developing decimal place-value system. It was only in the second half of the 19th century that the word "algorithm" acquired its present meaning, mostly through methods for solving recreational puzzles.

The situation changed dramatically in the 1940s and 1950s when algorithms for solving a wide range of problems began to emerge, partly through the requirements of warfare, and increasingly from the development of communication networks and the needs of the fledgling computer industry. In particular, linear programming surfaced at the end of World War II as an important area, both in practice (for solving practical optimization problems), and in theory (with the duality theorem as its outstanding minimax result). Previous workers on these problems included the French mathematician Joseph Fourier in 1824, the Hungarians Gyula (Julius) Farkas in1902, Dénes König in 1931, and Theodore S. Motzkin in 1934.

In this chapter we describe algorithms for specific practical problems in graph theory: the *minimum connector problem*, the *maximum flow problem*, the *shortest path problem*, and the *Chinese postman problem*. We then discuss the efficiency of algorithms in general, and the computational complexity of selected combinatorial problems. Complexity theory became a central area of graph theory in the second half of the 20th century; here, Jack Edmonds was an early pioneer, with an efficient algorithm for matching and with the conjecture that no efficient algorithm exists for the traveling salesman problem, now taking the more general form $\mathsf{P} \neq \mathsf{NP}$.

8.1 Algorithms

The earliest algorithms for graph theory may be those of the Viennese mathematician Carl Hierholzer and the Frenchman M. Fleury, who in 1873 and 1883 presented algorithms for tracing a route in an Eulerian graph (see Chapter 1). Around the same time, another Frenchman, Charles P. Trémaux, found an efficient algorithm for tracing a maze; it was presented in 1882 in the first volume of Édouard Lucas's four-volume work on recreational mathematics.[1] Better known, and simpler, is a maze-tracing algorithm of Gaston Tarry from 1895.[2] After presenting his general rule, Tarry gave a systematic method involving paper markers for carrying it out in practice.

Rule 8.1 (Tarry's rule). *Do not return from a junction along the passage which led to that junction for the first time unless you cannot do otherwise.*

Dénes König subsequently devoted a whole chapter of his influential textbook to the tracing of mazes.[3]

8.1.1 Two problems. One of the earliest general types of algorithm to be developed was also one of the most natural. This was the so-called *greedy algorithm*, which arose in connection with the following problem whose origins can be traced back to Czechoslovakia in the 1920s.

Problem 8.1 (Minimum connector problem). Given a number of cities connected by links that join cities in pairs, and given the distances between all pairs of cities, how can one connect the cities so that the total length of the connecting links is as small as possible?

As an example, consider the network of five cities in Figure 8.1. There are many ways of linking them, such as the network shown. Clearly the chosen links must form a tree, since otherwise any edge in a cycle could be removed, thereby reducing the total cost. But if there are n cities, then by Cayley's formula (see Chapter 7) there are n^{n-2} different trees joining them. This number increases rapidly with n, so which tree should be chosen? The greedy algorithm, which is described below and is easy to apply, always produces an optimal solution.

To apply the greedy algorithm, we select the shortest available edge at each stage, so long as its inclusion creates no cycle. In our given example, we choose links wx (length 2) and wz (length 3) first. We cannot then choose xz (length 4) because its addition would create the cycle $wxzw$, so we choose yz (length 5) instead. We cannot then choose wy or xy (length 6), since either would create a cycle, so we choose vw (length 7), thereby completing the tree. Thus the tree in Figure 8.1 is optimal, and its total length is 17.

The minimum connector problem seems to have been first posed and solved in 1926 by the Czech mathematician Otakar Borůvka, arising from a question concerning the electrification of Western Moravia. After presenting the problem more theoretically in a longer paper, "On a certain minimum problem", and deriving and proving

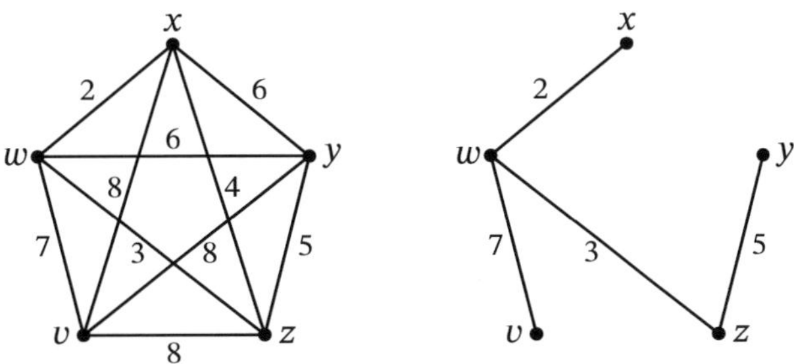

Figure 8.1. A minimum connector problem and its solution

an algorithm for solving it, he wrote a short and more accessible note in an engineering magazine,[4] giving the above version of the problem and solving an example with forty cities. He explained (without proof) how to construct the required tree by first linking each city to its nearest neighbor and then joining the fragments together. If the inter-city distances are all different, then the solution is unique. A translation and analysis of Borůvka's papers appears in a paper by Jaroslav Nešetřil, Eva Milkovă, and Helena Nešetřilová.[5]

Four years later, his compatriot Vojtěch Jarník presented a variation on the greedy algorithm. In a paper with the same title as Borůvka's first note,[6] Jarník applied his algorithm by choosing a city and then introducing each successive link in such a way that it joined a city previously considered to one not yet connected.

The writings of Borůvka and Jarník were in somewhat obscure publications, and in the 1950s the minimum connector problem was revisited on many occasions. In particular, two well-known and well-written papers appeared in American journals, by Joseph B. Kruskal and Robert C. Prim.[7] Kruskal's algorithm was essentially that of Borůvka's first note, while Prim used Jarník's approach and included a solution of a problem for forty-nine cities—the capital cities in the forty-eight contiguous states of the United States and the national capital (see Figure 8.2). Both authors acknowledged their debt to their Czech predecessors.

Around 1970, the algorithms used for solving such problems as the minimum connector problem were named "greedy" by Jack Edmonds,[8] who observed that the concept of a matroid, first introduced by Hassler Whitney in 1935, was a natural setting for such algorithms: indeed, a matroid is exactly the type of combinatorial structure for which the greedy algorithm works.

A related, although much more difficult, problem is to find the shortest tree connecting a number of given points in the plane if we are allowed to add extra points as intersections at which links can meet (see Figure 8.3). The problem seems to have originated with J. D. Gergonne[9] in 1810, and a number of 19th-century mathematicians worked on it, including the Swiss geometer Jakob Steiner who solved the problem for three points. The trees are now known as *Steiner trees*, and also arise in connection with the formation of soap bubbles. The problem for n points was discussed in 1934 by Vojtěch Jarník and Milos Kössler,[10] who proved the existence of such minimum-length trees and gave methods for finding them when the original points form a regular n-sided polygon.

Figure 8.2. The minimum connector for forty-nine cities

Another type of problem, which superficially resembles the minimum connector problem, is the "traveling salesman problem".

Problem 8.2 (Traveling salesman problem). Given a number of cities connected by links that join cities in pairs, and given the distances between all pairs of cities, what is the shortest cyclic route that passes through each city just once and returns to the starting point?

For example, consider again the five cities in Figure 8.1. Here, the optimum solution (found by trial and error) has total length 26, see Figure 8.4.

If there are n cities, then the number of cycles that pass through all the cities is $(n-1)!$ (or $\frac{1}{2}(n-1)!$ if travel is allowed in either direction). These numbers grow very quickly as n increases, and we might hope for an efficient algorithm that yields an optimum solution, rather like the greedy algorithm for the minimum connector problem. But no such efficient algorithm has ever been found. In 1967, Jack Edmonds conjectured that one cannot exist, and this is essentially the conjecture that $P \neq NP$ that

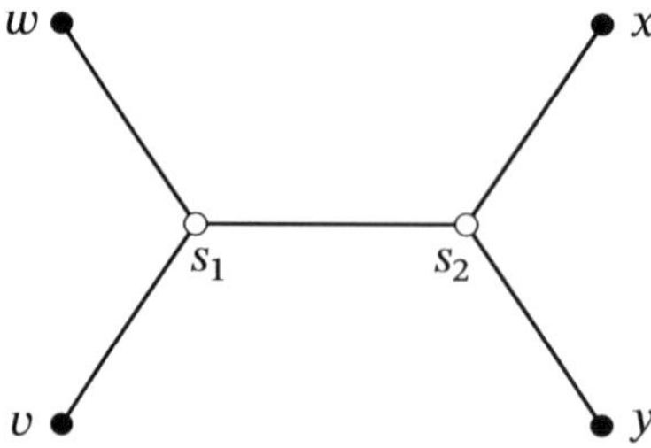

Figure 8.3. A Steiner tree having four cities and two additional sites

we discuss later in this chapter. To suggest the nonexistence of an efficient algorithm was a novel idea at the time.

The traveling salesman problem has been around since at least the 19th century, where it appeared in a sales manual of 1831. It was posed again by Karl Menger in Vienna in 1930, and soon afterwards at Harvard University by Hassler Whitney, who gave the problem its name. After being studied spasmodically in the 1940s it was brought into prominence in 1954 in an influential paper by George Dantzig, Delbert R. Fulkerson, and Selmer Johnson of the RAND Corporation, who solved the problem for the above example of forty-nine cities.[11] Since then, it has been solved for networks connecting many thousands of cities.

An efficient algorithm for solving the traveling salesman problem has not been discovered, and probably never will be, but there are heuristic algorithms that yield values that closely approximate the optimum one. Many books are now devoted to the problem, some of which are listed at the end of this chapter.

8.1.2 Matching and assignment algorithms. In matching and assignment problems, the aim is to assign various applicants in an efficient way to jobs for which they are qualified. As we saw in Chapter 5, matching problems are frequently represented by bipartite graphs, where the two partite sets correspond to the applicants and the jobs, and an edge joins an applicant-vertex to a job-vertex whenever the applicant is qualified for the job; the aim is to find a "maximum matching" in which the largest possible number of applicants are matched to suitable jobs. A simple example is given in Figure 8.5, where the five applicants above apply for the five jobs below. In this example, the only positions that applicants a, c, and d are qualified for are w and x, and so not everyone can get a job. However, four applicants can, as shown in the right-hand graph.

As we mentioned in Chapter 5, the matching problem later became known as the *marriage problem*, referring to boys and girls, with an edge representing a pair where the boy likes the girl. The task was to "marry" as many "liking pairs" as possible. Clearly, a necessary condition for all the boys to marry is that for each $k \leq n$ every group of k boys must collectively like at least k girls. The *König–Hall theorem* (named after Dénes König and Philip Hall, but also discovered by others in different contexts) states that this necessary condition is also sufficient for the existence of such a matching.

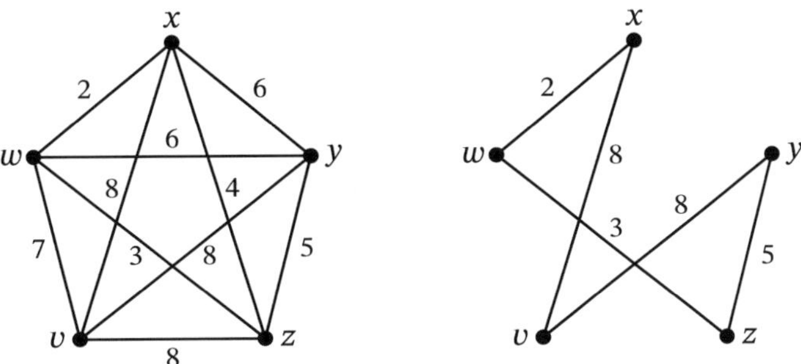

Figure 8.4. A traveling salesman problem and its solution

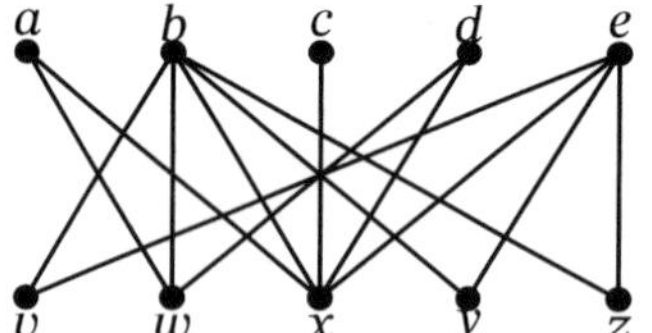
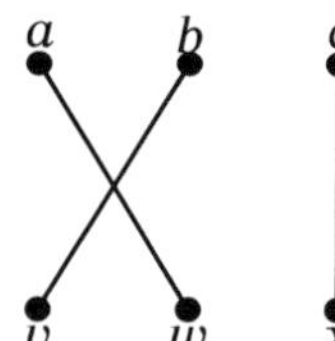
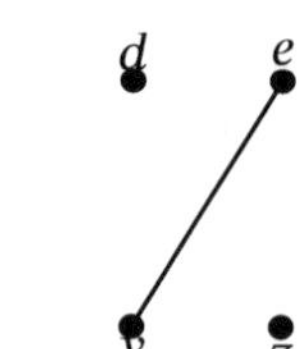

Figure 8.5. Optimal solution to a matching problem

Theorem 8.1 (König–Hall theorem). *In a bipartite graph with partite sets X and Y, there is a matching in which each vertex in X is matched to a unique vertex in Y if and only if each subset Z of vertices in X together has at least |Z| neighbors in Y.*

This result follows from König's minimax theorem (see Chapter 5). König proved it by presenting an efficient algorithm that, for a given bipartite graph, either finds a solution to marry all boys, or exhibits an explicit set of k boys who collectively like fewer than k girls.

In an assignment problem there may be several ways of allocating applicants to jobs, and the problem is then to determine which solution is the "best". Since certain applicants may be more suited to particular jobs, we assign to each edge of the bipartite graph a "cost" (or "ranking")—the lower the cost, the more suitable is that applicant for that job. The task is then to decide which of the possible maximum matchings has the lowest total cost.

Such problems were solved by several people, including the Hungarian mathematician Jenő (Eugene) Egerváry, who in 1931 extended König's minimax theorem to bipartite graphs with costs on the edges. In 1955, Harold Kuhn showed how the results of König and Egerváry give rise to a systematic method for solving assignment problems[12]—this method is now called the *"Hungarian algorithm"* in their honor. Kuhn's pioneering work was not expressed in the language of graph theory, but such a treatment now appears in many textbooks, and also in a survey by András Frank.[13]

8.1.3 Transportation and linear programming.

In 1941, the American mathematician and physicist Frank Hitchcock formulated the following general transportation problem.[14]

Problem 8.3 (Transportation problem). Several factories supply a product to a number of cities. Due to freight rates and other matters, the cost of shipping a ton of product to a particular city varies according to which factory supplies it, and also varies from city to city. What is the least expensive way in which to distribute the material?

This problem, in which the aim is to deliver a particular commodity through a network at minimum cost, is a development of the assignment problem for bipartite graphs, where not only do the edges have costs, but there are also supply requirements at each factory-vertex and demand requirements at each city-vertex. When these supplies and demands are all equal to 1, the transportation problem reduces to the assignment problem discussed above.

Such transportation problems can be traced back to 1784, when the French mathematician Gaspard Monge was concerned with a problem involving the transportation of quantities of earth.[15] Methods for solving it were developed by A. N. Tolstoĭ in Russia in the 1920s and 1930s, and then by Leonid Kantorovich and Tjalling Koopmans in the 1930s and 1940s, and by Hitchcock in his paper of 1941.

The contributions of these people led to the development of *linear programming*, which came to play an important role in the solution of the transportation and other problems. Although some of the subject's basic ideas on linear inequalities can be traced back to Joseph Fourier in the 1820s, Julius Farkas in 1902, and Theodore S. Motzkin in 1934,[16] the idea of linear programming in general was first developed in 1939 when Kantorovitch[17] presented such problem formulations, continuing his investigations throughout World War II. Concurrently, similar investigations were being pursued by Koopmans in an economic context, and in 1975 Kantorovich and Koopmans were awarded a Nobel Memorial Prize in Economics for their achievements.

Motivated by the wartime planning activities of the US Air Force, George Dantzig developed the basic theory of linear programming in 1946–47, introducing his *simplex method* for solving linear programming problems and illustrating its efficiency with an assignment problem involving 70 applicants and 70 jobs. A meeting between Dantzig and John von Neumann led to the latter's introduction of the important idea of *duality*, already implicit in the work of Fourier and Farkas. Further information can be found in Dantzig's 1963 book (see the Further reading) and in his historical reminiscences.[18]

8.1.4 Flows in networks. The theory of flows in capacitated networks was developed in the 1950s, mainly by Lester R. Ford Jr. and Delbert R. Fulkerson of the RAND Corporation. In 1956, they published a groundbreaking paper on network flows,[19] to be followed in 1962 by their classic book on the subject (see Further reading below). The basic network flow problem originated in 1955 in a query by T. E. Harris to the American Air Force (and downgraded to an unclassified problem only in 1999), concerning the Soviet railroad system.

Problem 8.4 (Network flow problem). Consider a network connecting two cities by way of a number of intermediate cities. If each link of the network has a number assigned to it, representing its capacity, find a maximum flow from one city to the other.

A simple example is given on the left in Figure 8.6, where we wish to find the maximum total flow of a commodity from the "source" s to the "sink" or "terminus" t without exceeding the capacity of any intermediate link. (The more general situation with several sources and sinks can easily be reduced to this simpler version.)

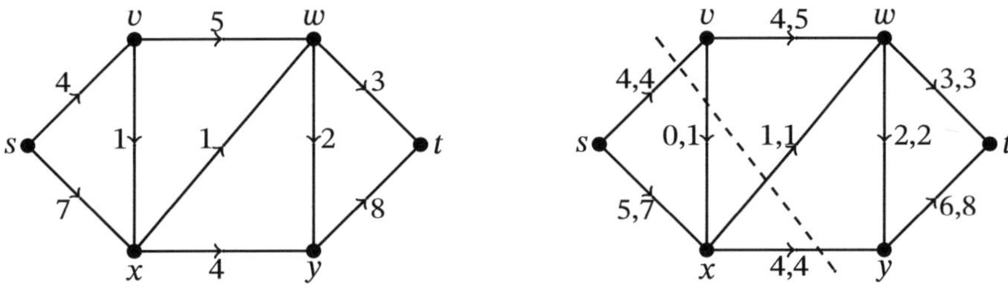

Figure 8.6. A network flow problem and its solution

To answer this, we look for paths from s to t along which the commodity can flow. One such path is $svwt$, along which we can send a flow of 3 without exceeding the capacity of any of its constituent arcs; the link wt is then "saturated" and can take no more flow. Other "flow-augmenting paths" are $sxyt$ (carrying a flow of 4) and $svxwyt$ (carrying a flow of 1). All directed paths from s to t then have a saturated arc, but the obtained total flow of 8 is not maximum because we can use $sxvwyt$ to increase the flow by 1 (note that $sxvwyt$ is not a directed path since the arc xv has the wrong direction; however, we increase the flow along xv by decreasing it on the directed edge vx). This gives a maximum total flow of 9, as shown on the right in Figure 8.6, where the first number next to each arc is the flow along it and the second is its capacity.

We note that the total flow into each intermediate vertex is equal to the total flow out of it, and that the arcs sv, xw, and xy are all saturated, while vx has no flow. We note also that if we delete the arcs sv, vx, xw, and xy (jointly carrying a flow of 9), then we split the network into two parts—one part containing the vertices s and x and the arc sx, and the other containing the remaining vertices and their linking arcs; such a set of arcs whose removal separates the network into two parts, one containing s and the other containing t, is called a *cut*, and the sum of the capacities of the directed arcs from the side containing s to the side containing t is the *capacity* of the cut.

We can now state the main result in the subject: Ford and Fulkerson's *max flow – min cut theorem*.

Theorem 8.2 (Max flow – min cut theorem). *In a capacitated network, the maximum flow from the source to the sink is equal to the minimum capacity of a cut separating the source from the sink.*

Ford and Fulkerson also pointed out deficiencies in the above method for finding flow-augmenting paths. The number of such paths used may depend on the capacities of the arcs, and when the capacities are not integers or rational numbers, the process might never stop. Both of these issues were remedied by Jack Edmonds and Richard Karp in 1972, by the use of flow-augmenting paths with as few edges as possible in each step.[20]

The max flow – min cut theorem is a particular case of linear programming duality. As we have mentioned, linear programming originated from specific transportations that arose between the two world wars, and its theory was developed after World War II. In the ensuing years, variations on the basic network problem were intensively studied—for example, where there are capacity restrictions on the vertices and costs on the edges to be taken into account.

A different type of flow problem is due to W. T. Tutte. A *k-flow* in an undirected graph is an assignment to each edge of a direction and a value from $1, 2, \ldots,$ $k - 1$ in such a way that, for each vertex v, the sum of the values of edges directed into v is equal to the sum of the values of the edges directed out of v. In 1954, Tutte proposed the following conjecture.[21] (Recall that the removal of a bridge from a connected graph separates it into two parts.)

Conjecture 8.1 (Tutte's 5-flow conjecture). *If a connected graph has no bridges, then it has a 5-flow.*

In the 1970s, François Jaeger and P. A. Kilpatrick proved this conjecture with 5 replaced by 8, and in 1981 Paul Seymour improved this to 6.[22] Tutte also conjectured that a connected graph with no separating set of one, two, or three edges has a 3-flow, and that a bridgeless graph that does not contain the Petersen graph as a minor has a 4-flow. This has developed into an active research area with connections to several other branches of graph theory, among them graph coloring. A monograph by C. Q. Zhang is mentioned in the Further reading section.

8.1.5 Path problems. The *shortest path problem* is that of determining a shortest route between two given cities in a network—for example, given a road map of the United States, what is the shortest highway route from San Diego, California, to Boston, Massachusetts? A simple example of a shortest-path problem is given on the left in Figure 8.7, where we wish to find a shortest path from the source s to the sink t.

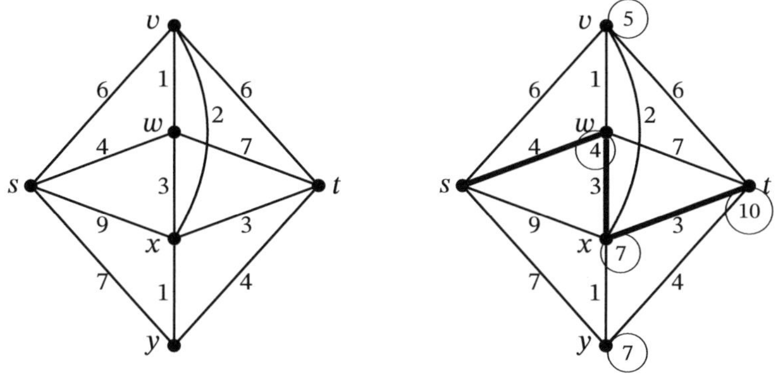

Figure 8.7. A shortest path problem and its solution

To find such a path we first identify the vertices that can be reached directly from s: these are v (at distance 6), w (at distance 4), x (at distance 9), and y (at distance 7). The smallest of these numbers is 4, and so we give w a permanent label of 4 (since there can be no shorter route from s to w). We also assign temporary labels of 6 to v, 9 to x, and 7 to y.

We next look at the vertices that are reached directly from w—these are v (at distance 1), x (at distance 3), and t (at distance 7). We can thus replace the temporary labels at v and x by 5 and 7, and give t a temporary label of 11. The temporary labels are now 5 at v, 7 at x, 7 at y, and 11 at t. The smallest of these is 5 at v, and so we make this label permanent. We next look at the vertices that are reached directly from v, and find that the smallest temporary labels are then at x and y, which are made permanent. Finally, looking at the vertices reached directly from x and y gives the sink t the new permanent label 10, as shown on the right in Figure 8.7. This is the shortest distance from s, and we find (on tracing back through the labels) that it arises from the path *swxt*.

Algorithms for finding such a shortest path were developed the 1950s, by Lester R. Ford Jr., Alfonso Shimbel, Richard Bellman, George Dantzig, Edward F. Moore, and others. Probably the best known of these algorithms, dating from 1959 and used in the above example, appeared in a short paper by the Dutch systems scientist Edsger Dijkstra,[23] who had applied it to a network of 64 places in the Netherlands, and who presented it together with his rediscovery of Jarník and Prim's algorithms for the minimum connector problem.

A related problem is the *longest path problem*, where we wish to find the longest path from a source to a sink. Such problems arise in *critical path analysis*, where we are concerned with the scheduling of tasks in a project (such as building a house or cooking a meal), and where certain tasks cannot be begun until others have been completed. This subject was developed in the 1950s by the U.S. Navy under the acronym of PERT (Program Evaluation and Review Technique) in connection with the development of the Polaris submarine weapon systems.

In a *Chinese postman problem* the aim is to find a shortest route that a postman can take in order to deliver a pile of letters to houses in a number of streets; the word "Chinese" refers to the original solver, Kwan Mei-Ko (Meigu Guan), and not to the problem.[24] The problem was given its name in 1965 by Edmonds, who presented a more efficient algorithm for its solution.[25]

If the network of roads containing the houses is an Eulerian graph, then the solution is clearly an Eulerian trail that includes each road just once; Fleury's algorithm can then be used to find a suitable route. But if it is not Eulerian, then the postman needs to traverse some roads more than once, and the problem is to minimize the amount of extra travel. This reduces to a weighted matching problem for the vertices of odd degree in the graph, where the weights are the lengths of the shortest paths connecting them. This results in an efficient algorithm, but surprisingly the problem becomes much more difficult if the network contains both one-way and two-way roads, as we shall mention in Section 8.2.

8.1.6 Searching graphs. In many practical problems, one needs to visit every vertex of a graph in a systematic way, and a number of searching algorithms have been developed. The best known of these are *depth-first search* and *breadth-first search*, both of which originated in the tracing of mazes—depth-first search by C. P. Trémaux in the 19th century, and breadth-first search by E. F. Moore in the 1950s.

In a depth-first search, we start by penetrating into the graph as deeply as possible before backtracking to explore other vertices. For example, if we start from vertex p in the tree of Figure 8.8, we might proceed to vertex q, and then directly to s and w (temporarily ignoring t, x, and z). We cannot then go any deeper, so we pick up x and z before backtracking to vertex p (picking up t on the way back to p). We then proceed along the other branch, visiting r, u, and y, picking up v on the way back to vertex p. This gives the vertex-ordering

$$p \to q \to s \to w \to x \to z \to t \to r \to u \to y \to v.$$

Another way of thinking about a depth-first search of a tree is to think of the tree as a brick wall and to go for a walk around it, always keeping the wall on one's right. The order in which one first visits the vertices yields a depth-first search. An example of the use of depth-first search in a general connected graph (yielding a tree containing all of the vertices of the graph), is to determine its cut-vertices.

In a breadth-first search the aim is to explore all nearby vertices before proceeding to more remote ones. For example, starting from vertex p in the tree in Figure 8.8, we first visit its neighbors q and r, then the neighbors of q (which are s and t) and the neighbors of r (which are u and v). Finally, we visit the neighbors of s and u (which are w, x, and y) and then the neighbor of x (which is z). This places the vertices of the

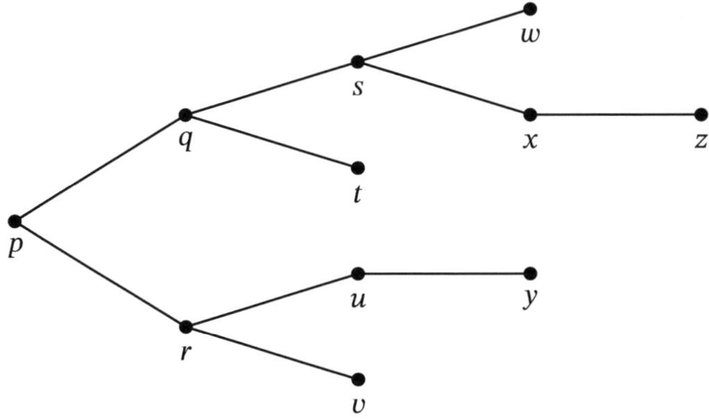

Figure 8.8. A depth-first search tree.

tree in alphabetical order,

$$p \to q \to r \to s \to t \to u \to v \to w \to x \to y \to z.$$

An example of the use of breadth-first search is in the solution of a shortest path problem, where we start with the vertices nearest to the source before moving outwards to more remote ones. This is used in the solution of network flow problems by Edmonds and Karp, where we seek flow-augmenting paths that reach from the source to the sink with as few edges as possible, as we explained above.

8.2 Complexity

Throughout the 1950s, as the rise in computer science was accelerating and as many new algorithms were being developed, it became ever more important to be able to distinguish between algorithms that are "efficient" and those that might take a long time to run.

This distinction had already been recognized by Thomas Malthus in 1798. In his *Essay on Population*, he contrasted the steady linear growth of food supplies with the exponential growth in population. He concluded that however we may cope in the short term, the exponential growth would inevitably win in the long term, and that there would be food shortages. In 2020, exponential growth became a familiar concept in connection with the spread of the COVID-19 virus.

Every algorithm has a *running time*, which may be the time that a computer needs to carry out all the necessary calculations, or the number of steps in these calculations. Each problem has an *input size*, which we call n—it may be the number of cities in a network or the number of digits in a large integer that we wish to factor—where the running time usually depends on the input size n.

8.2.1 The P = NP? problem. Particularly important, because they are the most efficient, are the *polynomial-time algorithms*, in which the maximum running time is bounded by a function that is proportional to a power of the input size, such as n^2 or n^5. The collection of all problem-types that can be solved by polynomial-time algorithms is called P.

Table 8.1. Running times

	$n = 10$	$n = 30$	$n = 50$
polynomial			
n	0.00001 seconds	0.00003 seconds	0.00005 seconds
n^2	0.0001 seconds	0.0009 seconds	0.0025 seconds
n^3	0.001 seconds	0.027 seconds	0.125 seconds
n^5	0.1 seconds	24.3 seconds	5.2 minutes
exponential			
2^n	0.001 seconds	17.9 minutes	35.7 years
3^n	0.059 seconds	6.5 years	2.3×10^{10} years

In contrast, many algorithms take more than polynomial time—for example, those that look through all 2^n subsets of a set of size n. In *exponential-time algorithms* the maximum running time may grow like (say) 2^n or 3^n. The collection of problem-types for which there is no polynomial-time algorithm is called *not*-P. In the following table of running times, we assume that our computer carries out one million calculations a second. As we can see, exponential growth algorithms generally take significantly longer than polynomial growth ones. Algorithms that run in polynomial time are generally thought to be "efficient", whereas those that do not run in polynomial time are considered "inefficient", certainly as the input size increases. Note, however, that a running time that is proportional to n^2 may in practice exceed one proportional to 2^n for small values of n if the constant of proportionality for the first is much greater than that for the second.

It was Edmonds who first promoted a theory of polynomial algorithms, when in 1965 he discovered such an algorithm for the maximum matching problem in a general graph (see Chapter 5), and observed:

> For practical purposes the difference between algebraic and exponential order is often more crucial than the difference between finite and nonfinite.

Several of the problems that we have mentioned here and in earlier chapters are in P: they have algorithms whose running time is at most some power of the input size—usually its square or cube. We say that the running time of a graph algorithm is $O(n^k)$ if it is at most a constant multiple of n^k, where n is the input size—usually the number of vertices. Examples for graphs with n vertices and m edges include

- the minimum connector problem, with running time $O(n^2)$;

- the maximum flow problem, with running time $O(n^3)$;

- the maximum matching problem, with running time $O(n^3)$;

- the shortest path problem, with running time $O(n^2)$;

- the Chinese postman problem, with running time $O(n^3)$;

- Fleury's algorithm for finding an Eulerian trail, with running time $O(m)$.

However, several related problems remain unsolved, such as:

Is there an algorithm with $O(nm)$ running time for the maximum flow problem?

Another type of problem with a polynomial algorithm is the problem of determining whether a given graph with n vertices is planar. This turns out to be a linear problem—its running time is $O(n)$—and was investigated by John Hopcroft and Robert Tarjan.[26] For their work in this and related areas, Hopcroft and Tarjan received the prestigious A. M. Turing Award from the Association of Computing Machinery in 1986.

A *decision problem* is a problem that has only the answer "yes" or "no": problems are often given in this form for easier comparison, and so P is the set of "easy" types of decision problems that can be solved with a polynomial-time algorithm. However, there are many types of problems for which no polynomial-time algorithm has ever been found, one of these being the traveling salesman problem. But for this problem, suppose that, for any suggested route-length k, you ask whether there is a solution that is less than k. If the answer is "yes", then there is a possible route and, given such a route, you can check in polynomial time that it is indeed a cyclic route with length less than k.

In general, the collection of such "nondeterministic polynomial-time problems"— those problems for which a "yes" answer has a solution (or a certain certificate) that can be checked to give the correct answer "yes" in polynomial time—is called NP. Clearly P is contained in NP, because the answer can already be found in polynomial time without a certificate (or in other words: with an empty certificate).

It follows from our above discussion that the traveling salesman problem is in NP. But is it in P? More generally, one can ask:

$$\text{Is } P = NP?$$

This seems very unlikely—indeed, few people believe that the answer to this question is "yes", because all problems for which a "yes" solution is easy to check would then also be easy to solve. But this has never been proved, and it is one of the major unsolved problems in mathematics today. The conjecture that $P \neq NP$ was essentially first formulated by Edmonds in 1967.

In 1965, Edmonds[27] also conjectured that $NP \cap co\text{-}NP = P$, where co-NP consists of those types of decision problems for which the answer "no" always has a certificate that can be checked efficiently. We can express this conjecture by saying "What is easy to verify is easy to find"—or, in other words:

> Types of problems for which answers have good mathematical characterizations are precisely those for which there are good algorithms for finding the answers.

This suggests a strong link between mathematics and computer science. An example of this is the perfect maximum matching problem for general graphs, with the good mathematicl characterization by Tutte (see Chapter 5). It shows that the problem is in $NP \cap co\text{-}NP$: if we ask for a perfect matching in a graph G then the answer "yes" can be verified by showing a perfect matching, and the answer "no" can be verified by showing as certificate a set S such that $G - S$ has more than $|S|$ odd connected components. This characterization of Tutte was then followed years later by the efficient algorithm of Edmonds. Another example is linear programming, with the duality theorem showing that it belongs to NP ∩ co-NP. It was not until 1979 that the first polynomial algorithm for linear programming was presented, by the Russian mathematician Leonid G. Khachiyan.[28]

Edmonds's two conjectures have been around for more than fifty years with almost no general progress in their solutions. Within the philosophy of mathematics and computer science, they tend to have taken the roles of generally accepted axioms that may not even be provable within our usual mathematical framework.

8.2.2 NP-complete problems. In 1971, Stephen Cook of the University of Toronto wrote a short but fundamental paper on the complexity of theorem-proving procedures,[29] in which he proved that a particular NP problem-type in mathematical logic, called the *satisfiability problem*, has the following remarkable property.

> If the satisfiability problem can be solved in polynomial time, then every problem in NP can be solved in polynomial time. More precisely, any problem in NP can be reduced in polynomial time to one that has the form of a satisfiability problem.

So if the satisfiability problem is in P, then everything else in NP is also in P, and we can deduce that P = NP. But if the satisfiability problem is not in P, then P ≠ NP. Thus, the whole "P = NP?" question relies on whether there is a polynomial algorithm for just a single problem-type.

A problem in NP is called NP-*complete* if its solution in polynomial time implies that every NP problem can be solved in polynomial time—so the satisfiability problem is NP-complete. Not every problem in NP needs to be NP-complete (unless P = NP), but the NP-complete problems certainly include many hundreds of problems, including the traveling salesman problem. So, if there were a polynomial algorithm for just one of these problems, then polynomial algorithms would exist for the whole collection, and P would be equal to NP. But if just one of them has no polynomial algorithm, then none of the others could have one either, and P would be different from NP.

The following are examples of NP-complete problems:

- the *traveling salesman problem*;

- the *longest path problem*;

- the *Hamiltonian cycle problem*: Does a given graph have a Hamiltonian cycle?

- the 3-*colorability problem*: Is a given graph 3-colorable?

- the *clique problem*: Does a given graph have a complete subgraph of at least a given size?

- the *Chinese postman problem* for a graph with both directed and undirected edges.

A decision problem whose status is unknown is the *graph isomorphism problem*, to decide whether or not two given graphs are isomorphic—that is, to decide whether there is a one-to-one mapping f from $V(G)$ onto $V(H)$ for which vw is an edge of G if and only if $f(v)f(w)$ is an edge of H. An important result in this direction is by Eugene M. Luks,[30] who showed in 1982 that graph isomorphism for graphs of bounded degree can be tested in polynomial time.

In this context we mention the problem of homomorphisms. A *homomorphism from G to H* is a function f from $V(G)$ into $V(H)$ that maps adjacent vertices in G to adjacent vertices in H: if vw is an edge of G then $f(v)f(w)$ is an edge of H. This generalizes the idea of graph coloring, because a homomorphism from G into the complete graph K_k is equivalent to a k-coloring of G. In 1990, Pavol Hell and Jaroslav Nešetřil proved that the problem of deciding whether an input graph G has a homomorphism into a fixed graph H is in P if H is bipartite, and is NP-complete otherwise.[31]

An early discussion of these issues can be found in a 1975 survey paper by Richard Karp.[32] Soon after, in 1979, Michael Garey and David Johnson published a classic book that contains a comprehensive list of more than 300 NP-complete problem-types (see the Further Reading section below).

8.3 Conclusion

As we have seen, graph theory has played an important role in the development of the theory of algorithms and their complexity, and conversely. Not only does algorithmic theory use examples from graph theory, but such examples and applications have been fundamental for creating some of its basic concepts, such as the classes P and NP.

The graph minor project of Robertson and Seymour, touched upon in Chapter 3, has deep algorithmic consequences. The Robertson–Seymour theorem, stated in Chapter 3, shows that any minor-closed family has a finite forbidden minor characterization, and the theory implies that recognizing whether a given graph is in the family is a problem in P. An important graph parameter, studied by Robertson and Seymour, is the so-called *tree-width*, measuring how close a graph is to having a tree-like structure. The class of graphs of bounded tree-width at most k is minor-closed. Minor theory is set to influence the development of graph theory for years to come.

Knowing whether P = NP is not just a theoretical matter—many NP-complete problems are of highly practical importance, and there is money at stake—including a prize of one million dollars offered by the Clay Mathematics Institute for settling the issue. However, little progress on this general problem has been made since the 1970s, and the statement that P ≠ NP seems almost to take the role as an axiom.

Further Reading

Other good sources for general combinatorial optimization problems and their solutions include

E. Lawler, *Combinatorial Optimization: Networks and Matroids*, Holt, Rinehart and Winston (1976);

D. J. A. Welsh, *Matroid Theory*, Academic Press (1976); reprinted by Dover (2010);

S. Even, *Graph Algorithms*, Computer Science Press (1979); 2nd edn., Cambridge Univ. Press (2012);

C. H. Papadimitriou and K. Steiglitz, *Combinatorial Optimization: Algorithms and Complexity*, Prentice Hall (1982); reprinted Dover (1998);

A. Schrijver, *Theory of Linear and Integer Programming*, Wiley (1986);

M. Grötschel, László Lovász, and Alexander Schrijver, *Geometric Algorithms and Combinatorial Optimization*, Springer (1988);

A. Schrijver, *Combinatorial Optimization: Polyhedra and Efficiency*, Wiley (1986);

J. Lee, *A First Course in Combinatorial Optimization*, Cambridge Univ. Press (2004);

J. Bang-Jensen and Gregory Gutin, *Digraphs: Theory, Algorithms and Applications*, Springer (2002);

A. Frank, *Connections in Combinatorial Optimization*, Oxford Univ. Press (2011).

The most comprehensive text for combinatorial optimization, with detailed historical notes, is

Alexander Schrijver, *Combinatorial Optimization; Polyhedra and Efficiency* (3 vols.), Springer (2003).

The "bible" of linear programming, rich in insight and coverage, is

G. B. Dantzig, *Linear Programming and Extensions*, Princeton Univ. Press and RAND Corp. (1963); reprinted by Princeton Univ. Press (1998).

The standard monograph on P, NP, and NP-complete problems is

M. R. Garey and D. S. Johnson, *Computers and Intractability: A Guide to the Theory of NP-completeness*, Freeman (1979).

Books on the traveling salesman problem include

E. L. Lawler, J. K. Lenstra, A. H. G. Rinnooy Kan, and D. B. Shmoys (eds.), *The Traveling Salesman Problem: A Guided Tour of Combinatorial Optimization*, Wiley (1985);

D. L. Applegate, R. E. Bixby, V. Chvátal, and W. J. Cook, *The Traveling Salesman Problem*, Princeton Univ. Press (2007);

G. Gutin and A. P. Punnen, *The Traveling Salesman Problem and its Variations*, Springer (2007);

W. J. Cook, *In Pursuit of the Traveling Salesman: Mathematics at the Limits of Computation*, Princeton Univ. Press (2012).

Other books on more specific topics are

L. R. Ford, Jr. and D. R. Fulkerson, *Flows in Networks*, Princeton Univ. Press (1962);

L. Lovász and M. D. Plummer, *Matching Theory*, North-Holland (1986); reprinted by Elsevier (2009);

C. Q. Zhang, *Integer Flows and Cycle Covers of Graphs*, Marcel Dekker (1997).

A general volume containing around forty papers by leading scholars is

Martin Grötschel (ed.), *Optimization Stories*, Documenta Mathematica Extra Series, EMS (2012).

Notes and References

[1]Some writings from the late 1800s are
- C. Hierholzer, Ueber die Möglichkeit, einen Linienzug ohne Wiederholung und ohne Unterbrechnung zu umfahren, *Math. Ann.* 6 (1873), 30–32;
- M. Fleury, Deux problèmes de géométrie de situation, *J. Math. Élement.* 2 (1883), 257–261;
- É. Lucas, *Récréations Mathématiques*, Vol. 1, Gauthier-Villars, Paris (1882).

[2]G. Tarry, Le problème des labyrinthes, *Nouv. Ann. Math.* (3) 14 (1895), 187–190.

[3]D. König, *Theorie der endlichen und unendlichen Graphen*, Teubner (1936, 1986); English transl. *Theory of Finite and Infinite Graphs*, Birkhäuser (1990). See the Further Reading section in Chapter 1.

[4]O. Borůvka, O jistém problému minimálním, English transl. On a certain minimal problem, *Práce Moravské Přirodovědecké Spolecnosti (Brno), Acta Soc. Scient. Natur. Moraviae* 3 (1926), 37–58 and O. Borůvka, A contribution to the solution of a problem of economical construction of electrical networks [in Czech], *Elektronický Obzor* 15 (1926), 153–154.

[5]J. Nešetřil, E. Milková, and H. Nešetřilová, Otakar Borůvka, On minimum spanning tree problem (Translation of both 1926 papers; comments and history), *Discrete Math.* 231 (2001), 3–36.

[6]V. Jarník, O jistém problému minimálním, English transl. On a certain minimal problem, *Práce Moravské Přirodovědecké Spolecnosti (Brno), (Acta Soc. Scient. Natur. Moraviae)* 6 (1930), 57–63.

[7]J. B. Kruskal, Jr., On the shortest spanning subtree of a graph and the traveling salesman problem, *Proc. Amer. Math. Soc.* 7 (1956), 48–50;
R. C. Prim, Shortest connection networks and some generalizations, *Bell Systems Tech. J.* 36 (1957), 1389–1401.

[8]J. Edmonds, Matroids and the greedy algorithm, *Math. Program.* 1 (1971), 127–136.

[9]J. D. Gergonne, Solutions purement géométriques des problèmes de minimis proposés aux pages 196, 232 et 292 de ce volume, et de divers autres problèmes analogues, *Annales de math. pures et appliquées* 1 (1810), 375–384.

[10]V. Jarník and M. Kössler, O minimálnich grafech obsahujicich n daných bodů, English transl. On minimal graphs containing n given points, *Časopis Pěst. Mat. a Fysiky* 63 (1934), 223–235.

[11]G. Dantzig, R. Fulkerson, and S. Johnson, Solution of a large-scale traveling-salesman problem, *Oper. Res.* 2 (1954), 393–410. This paper introduced the "cutting plane method" in integer programming that has since been widely used to solve a range of combinatorial optimization problems.

[12]H. W. Kuhn, The Hungarian method for the assignment problem, *Naval Res. Logistics* 2 (1955), 83–97.

[13]A. Frank, On Kuhn's Hungarian method—A tribute from Hungary, *Naval Res. Logistics* 52 (2005), 2–5. This paper was written for a celebration by the Hungarian Academy of Sciences in Budapest of the 50th anniversary of the Hungarian method, with Kuhn as a main speaker.

[14]F. L. Hitchcock, The distribution of a product from several sources to numerous localities, *J. Math. Phys.* 20 (1941), 224–230.

[15]G. Monge, Mémoire sur la théorie des déblais et des remblais, *Histoire de l'Académie Royale des Sciences avec Mémoire Math. et Physique*, 2ème partie, Paris (1784), 34–38 (*Histoire*), 666–704 (*Mémoire*).

[16]The first mention of Fourier's method of elimination appeared in pages 47–55 of J. B. J. Fourier, *Histoire de l'Academie Royale des Sciences de l'Institut de France (Partie Mathématiques)* (1824), 1–91. A further description was presented in J. Fourier, *Oeuvres II* (1826), 317–328. The main contributions by Farkas and Motzkin are J. Farkas, Über die Theorie der einfachen Ungleichungen, *J. Reine Angew. Math.* 124 (1902), 1–24 and T. S. Motzkin, *Beiträge zur Theorie der linearen Ungleichungen*, Dissertation, University of Basel (1936).

[17]An early publication on linear programming was
L. Kantorovich, *The Mathematical Method of Production Planning and Organization* (Russian), Leningrad State Univ. (1939).

[18]G. Dantzig, Reminiscences about the origins of linear programming, *Oper. Res. Letters* 1 (1982), 43–48.

[19]L. R. Ford, Jr. and D. R. Fulkerson, Maximal flow through a network, *Canad. J. Math.* 8 (1956), 399–404.

[20]J. Edmonds and R. M. Karp, Theoretical improvements in algorithmic efficiency for network flow problems, *J. Assoc. Comput. Mach.* 19 (1972), 248–264.

[21]W. T. Tutte, A contribution to the theory of chromatic polynomials, *Canad. J. Math.* 6 (1954), 80–91.

[22]P. D. Seymour, Nowhere-zero 6-flows, *J. Combin. Theory (B)* 30 (1981), 130–135.

[23]E. W. Dijkstra (1959), A note on two problems in connexion with graphs, *Numerische Math.* 1 (1959), 269–271.

[24]M.-K. Kwan (= M. Guan), Graphic programming using odd or even points, *Chinese Math.* 1 (1962), 273–277.

[25]J. Edmonds, The Chinese postman's problem, *Bull. Oper. Res. Soc. Amer.* 13 (1965), B-73.

[26]J. Hopcroft and R. Tarjan (1974), Efficient planarity testing, *J. Assoc. Comput. Mach.* 21 (1974), 549–568.

[27]J. Edmonds, Minimum partition of a matroid into independent subsets, *J. Res. Nat. Bur. Standards* 69B (1965), 67–72.

[28]L. G. Khachiyan, A polynomial algorithm in linear programming, *Doklady Akademii Nauk SSSR* 244 (1979), 1093–1096 (Russian); English transl. *Sovjet Mathematics Doklady* 20 (1979), 191–194. Reactions to Khachiyan's paper are described in E. L. Lawler, The great mathematical sputnik of 1979, *The Sciences* 20 (1980), 12–15. Khachiyan's paper was quickly brought to general attention in a lecture by Lovász at the International Symposium on Mathematical Programming in Montreal in August 1979.

[29]S. A. Cook, The complexity of theorem-proving procedures, *Proc. 3rd Annual ACM Symposium in the Theory of Computing*, Shaker Heights, Ohio, 1971, ACM (1971), 151–158. This paper gave precise definitions of P, NP, co-NP and NP-completeness, based on Turing machines and the precise definition of the concept algorithm. Thus, the conjectures $P \neq NP$ and $NP \cap co\text{-}NP = P$ became precise mathematical questions, rather than meta-mathematical statements.

[30]E. M. Luks, Isomorphism of graphs of bounded valence can be tested in polynomial time, *J. Comput. System Sci.* 25 (1982), 42–65.

[31]P. Hell and J. Nešetřil, On the complexity of H-coloring, *J. Combin. Theory (B)* 48 (1990), 92–110.

[32]R. M. Karp, On the computational complexity of combinatorial problems, *Networks* 5 (1975), 45–68.

Credits

Cover Photos:
- The photo of König is courtesy of the János Bolyai Mathematical Society.
- The photo of Kuratowski is from Wikimedia Commons: `https://commons.wikimedia.org/wiki/Category:Kazimierz_Kuratowski`.
- The photo of Erdős is a private photo due to the authors.
- The photo of Tutte is courtesy of the University of Waterloo Archives.
- The photo of Whitney is from Wikimedia Commons: `https://commons.wikimedia.org/wiki/Category:Hassler_Whitney_(mathematician)`.

Figure 1.5: Wikimedia commons, "The seven bridges of Königsberg". A copy of Euler's original drawing from 1736. `https://commons.wikimedia.org/wiki/File:The_Seven_Bridges_of_Königsberg,_Fig._1.png` (PD)

Figures 1.13, 2.1, 2.3, 2.5, and 2.6: Adapted from *Four Colors Suffice: How the Map Problem Was Solved - Revised Color Edition* by Robin Wilson, with a new foreword by Ian Stewart. Copyright ©2002, 2014 by Robin Wilson. Reprinted by permission of Princeton University Press.

Figure 3.8: Wikimedia Commons, "Simple Torus", by YassineMrabet, (CC-BY-SA) `https://commons.wikimedia.org/wiki/File:Simple_Torus.svg`

Figure 4.8: Wikimedia Commons, "Hoffman-Singleton graph", (CC-BY-SA) `https://commons.wikimedia.org/wiki/File:Hoffman-Singleton_graph.svg`

Figure 5.1: Wikimedia Commons, "Tutte graph", (CC-BY-SA) `https://commons.wikimedia.org/wiki/File:Tutte_graph.svg`

Figures 3.9 and 3.16 Lowell W. Beineke and Robin J. Wilson (eds.), Graph Connections, Oxford Science Publications, 1997: Chapter 11, "Topology", by Lowell Beineke. Reprinted by permission of Oxford University Press.

Figures 3.13, 3.14, and 3.15: Reproduced with permission from L. W. Beineke and R. J. Wilson (eds.), *Topics in Topological Graph Theory*, Cambridge University Press, 2009, chapter by the editors.

Figure 8.2: The original figure is from R. C. Prim, *Shortest connection networks and some generalizations*, Bell System Technical Journal 36 (1957), 1389–1401. This journal existed from 1923–1957, and is available from the webpages of IEEE. This copy is from `https://www.biodiversitylibrary.org/page/17281161`. There is no copyright, as the web-page explains: "BHL offers a wide range of free tools and services to support the use and re-use of our collections and data."

Index